Synthese Library

Studies in Epistemology, Logic, Methodology, and Philosophy of Science

Volume 520

The aim of *Synthese Library* is to provide a forum for the best current work in the methodology and philosophy of science and in epistemology, all broadly understood. A wide variety of different approaches have traditionally been represented in the Library, and every effort is made to maintain this variety, not for its own sake, but because we believe that there are many fruitful and illuminating approaches to the philosophy of science and related disciplines.

Special attention is paid to methodological studies which illustrate the interplay of empirical and philosophical viewpoints and to contributions to the formal (logical, set-theoretical, mathematical, information-theoretical, decision-theoretical, etc.) methodology of empirical sciences. Likewise, the applications of logical methods to epistemology as well as philosophically and methodologically relevant studies in logic are strongly encouraged. The emphasis on logic will be tempered by interest in the psychological, historical, and sociological aspects of science. In addition to monographs *Synthese Library* publishes thematically unified anthologies and edited volumes with a well-defined topical focus inside the aim and scope of the book series. The contributions in the volumes are expected to be focused and structurally organized in accordance with the central theme(s), and should be tied together by an extensive editorial introduction or set of introductions if the volume is divided into parts. An extensive bibliography and index are mandatory.

Seungbae Park

Induction, Science, and Morality

Springer

Seungbae Park
School of Liberal Arts
Ulsan National Institute of Science and Technology
Ulsan, Korea (Republic of)

ISSN 0166-6991 ISSN 2542-8292 (electronic)
Synthese Library
ISBN 978-3-032-00031-6 ISBN 978-3-032-00032-3 (eBook)
https://doi.org/10.1007/978-3-032-00032-3

This Springer imprint is published by the registered company Springer Nature Switzerland AG
The registered company address is: Gewerbestrasse 11, 6330 Cham, Switzerland

Preface

This book is intended to contribute to the four debates: the debate on the problem of induction, the scientific realism debate, the mathematical realism debate, and the moral realism debate. It aims to refute Hume's argument against inductions and to defend the three positions: scientific realism, mathematical antirealism, and moral antirealism. Many philosophers would find these aims disagreeable or unachievable. On their account, the problem of induction is not solvable, and scientific, mathematical, and moral realism are epistemically on a par.

Whoever might be right, the four debates are intricately connected with one another, as will become clear in this book. For example, some scientific antirealists retreat to inductive skepticism to avoid criticisms against scientific antirealism. Some moral realists rely on scientific and mathematical realism to establish moral realism. Inductive skeptics may raise the problem of induction not only against scientific realism but also against moral realism. A new idea in one debate can have ramifications for other debates.

I could not have written this book without my intellectual debt to my graduate advisor, Dr. Richard Healey, at the University of Arizona. Philosophy of science that I learned from him is the very foundation of this book. I still remember that he said to me that when writing a paper, I should always think as if my philosophical opponent is right next to me. I understood why he said so during my career of publishing papers. I wish that I could go back to the good old days when I took graduate courses from him and TA-ed for him.

This book also benefited greatly from the graduate courses that I took at the University of Arizona. They are the metaethics course taught by Dr. Ron Milo, the epistemology and philosophy mind courses taught by Dr. Alvin Goldman, and the history of modern philosophy course co-taught by Dr. David Owen and Dr. Keith Lehrer. Many ideas discussed in those great courses are woven into this book. I remember, for example, that I wrote a term paper on Hume's problem of induction in the history of modern philosophy course.

I received valuable comments on earlier drafts of this book from Dr. Dave Truncellito at Language Scholars, Dr. Yafeng Shan at the Hong Kong University of Science and Technology, Dr. Byeong D. Lee at Sungkyunkwan University, Dr. Shin

Kim at Hankuk University of Foreign Studies, and Dr. Kane Baker at University of Exeter. I highly recommend Dr. Baker's video clips in YouTube to the readers of this book. I used to listen to his videos while dropping my kids off at school and picking them up from school.

Five chapters of this book improved significantly after their earlier versions were presented in the conferences as indicated in the following table:

	Chapters	Organizers	Dates
1	Chapter 1	The Korean Society for Analytic Philosophy	August 28, 2022
2	Chapter 4	The Korean Society for Philosophy of Science	June 30, 2022
3	Chapter 11	The Korean Society for Analytic Philosophy	December 10, 2022
4	Chapter 12	The Korean Society for Analytic Philosophy and the Korean Association for Logic	February 10, 2023
5	Chapter 14	The Korean Society for Analytic Philosophy	May 13, 2023

I thank the organizers of the conferences for giving me the chance to present the drafts of the chapters and the participants in the conferences for their useful comments.

The manuscript of this book is completed while I visit University of California–San Diego as a visiting scholar from 2023 to 2024. I am grateful to my sponsor faculty, Dr. Gila Sher, and the department head, Dr. Dana Nelkin, for their hospitality. I previously visited this department for a year from 2015 to 2016 when Dr. Craig Callender was the department head. I liked this department and my family liked San Diego so much that we came back.

My gratitude also goes to my wife (Hyegyoung Gong) and four children (Jihye Park, Jeongae Park, Suzie Park, and Eugene Park). Without them, this book could not exist. Their very existence is the source of delight and comfort to me. This book will give my children and their children an idea of how I lived my life and how they can live their lives.

Finally, I thank Dr. Otávio Bueno, the Editor-in-Chief, for inviting me to contribute to Synthese Library. Thinkers feel most honored when their thoughts are appreciated by their fellow thinkers. I hope that this book will turn out to be useful to Synthese Library as well as to readers.

Ulsan, The Republic of Korea Seungbae Park

Funding Acknowledgment

This work was supported by the Ministry of Education of the Republic of Korea and the National Research Foundation of Korea (NRF-2018S1A5A2A01039606).

Contents

Chapter 1
Critiques of Hume's Argument Against Inductions

Abstract I raise the following six objections to Hume's argument against inductions. (1) The faculty of reason and experience might jointly justify the conclusions of inductions. Accordingly, just because the conclusions of inductions are neither justified a priori nor a posteriori, it does not follow that they are not justified. (2) Inductions are required for the uniformity principle to be justified rather than the other way around. (3) It is self-defeating for Hume to use inductions to show that inductions are not justifiable. (4) The faculty of reason and experience might jointly justify the uniformity principle. Consequently, just because the uniformity principle is neither justified a priori nor a posteriori, it does not follow that it is not justified. (5) It is in principle impossible to construct an argument that would show that inductions are justified or unjustified. (6) Hume's argument against inductions backfires on his positive account of why we make predictions.

Keywords Counter-induction · Induction · Hume · Problem of induction · Uniformity principle

1.1 Introduction

Inductive inferences are widely used in our daily lives and in science. It is an inductive inference, for example, that since observed gold bars are yellow, unobserved gold bars are also yellow. It is also an inductive inference that since overeating and obesity were conjoined in the past, they will also be conjoined in the future. In such inferences, even if premises are true, conclusions might be false.

David Hume (1993, pp. 15–25) develops the celebrated philosophical problem in the history of philosophy called "the problem of induction," which holds that inductions are not justifiable, i.e., that "there can be no justification of induction" (Norton, 2003, p. 666). He has constructed a sophisticated argument against inductions to show that they are not justifiable. This chapter aims to analyze and criticize it.

This chapter is organized as follows. In Sect. 1.2, I analyze Hume's argument against inductions, laying out its premises and conclusion. In Sect. 1.3, I raise the

S. Park, *Induction, Science, and Morality*, Synthese Library 520,
https://doi.org/10.1007/978-3-032-00032-3_1

following six objections to it. (1) The faculty of reason and experience might jointly justify the conclusions of inductions. Accordingly, just because the conclusions of inductions are neither justified a priori nor a posteriori, it does not follow that they are not justified. (2) Inductions are required for the uniformity principle to be justified rather than the other way around. (3) It is self-defeating for Hume to use inductions to show that inductions are not justifiable. (4) The faculty of reason and experience might jointly justify the uniformity principle. Consequently, just because the uniformity principle is neither justified a priori nor a posteriori, it does not follow that it is not justified. (5) It is in principle impossible to construct an argument that would show that inductions are justified or unjustified. (6) Hume's argument against inductions backfires on his positive account of why we make predictions. In Sect. 1.4, I reply to a possible objection that induction can be invalidated by counter-induction.

My approach to the problem of induction is radically distinct from those of other philosophers in the literature. Ever since Hume advanced the problem of induction, many philosophers have been attempting to solve it, i.e., to justify inductions. They have been presupposing all along that Hume's argument against inductions is convincing. However, I reject their presupposition, and I do not attempt to solve the problem of induction, i.e., to show that inductions are justified. I instead attempt to refute Hume's argument against inductions, thereby showing that it fails to establish that inductions are not justifiable.

This chapter will disappoint those who expect a brilliant idea that would justify inductions, but will reward those who wonder what Hume's argument against inductions is, whether its premises are true, whether the inference from its premises to its conclusion meshes well with what its conclusion asserts, whether inductions are justifiable, and whether Hume's position is coherent.

1.2 Hume's Argument Against Inductions

In this section, I analyze Hume's argument against inductions, making it explicit what its premises and conclusion are.

According to Hume, the premises of inductions are justified by experience. He writes, "As to past *Experience*, it can be allowed to give *direct* and *certain* information of those precise objects only" (Hume, 1993, p. 21). For example, experience justifies the premise of the inductive inference that since observed gold bars are yellow, unobserved gold bars are also yellow, and the premise of the inductive inference that overeating and obesity were conjoined in the past, they will also be conjoined in the future.

How about the conclusions of inductions? According to Hume, they are neither justified a priori nor a posteriori. They are not justified a priori because their negations "can never imply a contradiction" (Hume, 1993, p. 15). For example, it is not a contradiction to say that unobserved gold bars are not yellow. Nor is it a contradiction to say that overeating and obesity will not be conjoined in the future. Therefore, the faculty of reason does not justify the conclusions of inductions.

The conclusions of inductions are not justified a posteriori either because cause and effect are not necessarily connected with each other, i.e., even if cause occurs, "consequence seems no wise necessary" (Hume, 1993, p. 21). For example, there is no necessary connection between the property of being a gold bar and the property of being yellow. Nor is there necessary connection between overeating and obesity. Therefore, experience does not justify the conclusions of inductions.

Let me make a quick comment on Hume's foregoing argument against the conclusions of inductions. To say that the conclusions of inductions are not justified a priori because their negations do not involve contradictions implies that they are not justified a priori because they might be false. To say that the conclusions of inductions are not justified a posteriori because there is no necessary connection between properties or between events implies that they are not justified a posteriori because they might be false. Thus, the faculty of reason does not justify the conclusions of inductions ultimately because they might be false, and that experience does not justify the conclusions of inductions ultimately because they might be false. To put it otherwise, the possibility that the conclusions of inductions might be false is ultimately what grounds Hume's contention that neither the faculty of reason nor experience justifies inductions.

Inductions would be justified, according to Hume, if they were supplied with a proposition that connects the premises with the conclusions. He says, "There is required a medium, which may enable the mind to draw such an inference" (Hume, 1993, p. 22). The required medium is the uniformity principle, according to which nature is uniform. This idea can be illustrated by the following two arguments:

(I)
Observed gold bars are yellow.
∴ Unobserved gold bars are yellow.

(U)
Observed gold bars are yellow.
Nature is uniform.
∴ Unobserved gold bars are yellow.

On Hume's account, the inference from the premise to the conclusion of (I) is not justified, while the inductive inference from the two premises to the conclusion of (U) is justified. To put it differently, the inference from observed gold bars to unobserved ones is not legitimate without the uniformity principle.

Is the uniformity principle justified? Hume claims that it is not justified a priori because "it implies no contradiction, that the course of nature may change" (Hume, 1993, p. 22). Neither is it justified a posteriori because to appeal to an argument from experience to justify the uniformity principle "must be evidently going in a circle, and taking for granted, which is the very point in question" (Hume, 1993, p. 23). In other words, it is impossible "that any arguments from experience can prove this resemblance of the past to the future; since all these arguments are founded on the supposition of that resemblance" (Hume, 1993, p. 24). To state it otherwise, since the uniformity principle is required for all inductions, it is circular to provide an induction to justify the uniformity principle. Therefore, the uniformity principle is neither justified a priori nor a posteriori.

Hume's argument against inductions recounted so far can be summed up in the following standard form:

Inductions are neither justified a priori nor a posteriori.
The uniformity principle is required for inductions to be justified.
The uniformity principle is neither justified a priori nor a posteriori.
∴ Inductions are not justifiable.

Keep this standard form in mind, as I will refer back to the premises and the conclusion in Sect. 1.3 below, where I criticize each premise as well as the inference from the premises to the conclusion.

1.3 Critical Responses

1.3.1 The First Premise

In this subsection, I criticize the first premise of Hume's argument against inductions that inductions are neither justified a priori nor a posteriori.

As noted in Sect. 1.2 above, Hume considers the possibility that the faculty of reason might justify the conclusions of inductions, and then rejects it on the ground that negating the conclusions of inductions does not give rise to contradictions, i.e., ultimately on the grounds that they might be false. He also considers the possibility that experience might justify the conclusions of inductions, and then rejects it on the grounds that cause and effect are not necessarily connected with each other, i.e., ultimately on the grounds that the conclusions of inductions might be false. Surprisingly, he does not consider the possibility that the faculty of reason and experience might jointly justify the conclusions of inductions.

Why is it more plausible to say that the faculty of reason and experience jointly justify the conclusions of inductions than to say that they separately justify the conclusions of inductions? Consider the induction that since observed gold bars are yellow, unobserved gold bars are also yellow. We would not form the premise in our minds without the observation of some gold bars. Thus, the premise is a product of experience. In addition, even if we formed the premise, we would not form the conclusion without the inference from the premise to the conclusion. The inference is a product of the operation of the mind. Thus, the conclusion is a joint product of the mind and experience. Therefore, the conclusions of inductions are more likely to be jointly justified by the faculty of reason and by experience than by the faculty of reason alone or by experience alone.

The first premise of Hume's argument against inductions is built upon the demanding conception of justification according to which if a belief might be false, it is not justified, i.e., a justified belief is guaranteed to be true. Contemporary epistemologists (Goldman & Beddor, 2021) reject this demanding conception of justification and accept the undemanding conception of justification according to which a justified belief is likely to be true, and thus that a justified belief might be false. On the undemanding conception of justification, you cannot say that the conclusions of inductions are not justified merely on the grounds that negating them does not give rise to contradictions and merely on the grounds that cause and effect are not necessarily connected with each other, i.e., ultimately on the grounds that the conclusions of inductions might be false.

Relatedly, Hume's view that the conclusions of inductions are not justified requires not the mere possibility statement that they might be false but rather a probabilistic statement that they are likely to be false. Establishing the probabilistic statement, in turn, requires that Hume randomly select a large number of inductions from the population of all inductions, and then show that the conclusions of the selected inductions are mostly false. If he shows that they are mostly false, inductivists, those who have embraced inductions, would happily give up inductions.

Let me turn to an internal conflict in Hume's position. On the one hand, he claims that the conclusions of inductions are not justified because they might be false. On the other hand, he claims that experience justifies the premises of inductions. In my view, just as the conclusions might be false, so the premises might be false. Consider, for example, the premise that observed gold bars are yellow. It might be evil demon, not the world, that causes the perception of yellow gold bars in our minds. Therefore, the premise might be false, and hence it is not justified.

If you are not convinced of this conclusion, I invite you to consider the following Humean argument against the premises of inductions. The premises of inductions are not justified a priori because their negations do not commit contradictions. They are not justified a posteriori either because there is no necessary connection between the perception of objects and objects, i.e., because objects might not exist although the perception of them exists in our minds.

The premises of inductions require a prior belief that I call "the causation principle" according to which the world causes experience in our minds, i.e., without the causation principle, the premises of inductions are not justified. However, the causation principle is not justified a priori because it is not a contradiction to say that the world does not cause experience in our minds. Neither is it justified a posteriori because it is circular to appeal to experience to justify it. Therefore, the premises of inductions are not justifiable, contrary to what Hume assumes.

Hume might reply that the causation principle better explains the existence of experience in our minds than the Cartesian hypothesis that the evil demon causes experience in our minds. In consequence, the causation principle is justified while the Cartesian hypothesis is not. Unfortunately, Hume cannot avail himself of this argument for the causation principle. After all, the argument exemplifies an inductive inference called "inference to the best explanation" (IBE). IBE is epistemically worse than enumerative induction. Enumerative induction is an inference from observed objects to unobserved objects which are observable. By contrast, IBE is an inference from observed objects not only to unobserved objects which are observable but also to unobserved objects which are unobservable, such as quarks and electrons. If enumerative induction is not justifiable, IBE is a fortiori not justifiable.

An interesting consequence follows from the foregoing Cartesian objection to Hume's contention that experience justifies the premises of inductions. It is arbitrary for Hume to say that experience justifies the premises of inductions, but not the conclusions of inductions. The perception of overeating and obesity is not necessarily connected with overeating and obesity any more than overeating is necessarily connected with obesity. In other words, obesity might not exist in the future although overeating exists. Similarly, none of overeating and obesity might have existed although the perception of them existed in our minds. However, Hume holds a double standard toward these two possibilities. He claims that experience justifies the premise that overeating coexisted with obesity *despite* the possibility that none of them might have existed, but that experience does not justify the conclusion that overeating will coexist with obesity in the future *because* obesity might not exist in the future although overeating exists.

Hume has the burden of explaining why he takes the possibility seriously that obesity might not exist in the future although overeating exists, but not the possibility that none of overeating and obesity might have existed although the perception of them existed. Why is it that the former possibility prevents experience from justifying the conclusion, but that the latter possibility does not prevent experience from justifying the premise? In other words, why is it that experience does not justify the conclusion because of the possibility that overeating and obesity might not be conjoined with each other in the future, but that experience justifies the premise in spite of the possibility that the perception of overeating and obesity might not be conjoined with overeating and obesity? It is beyond my imagination how Hume would justify his double standard toward the two possibilities.

To sum up, the faculty of reason and experience might jointly justify the conclusions of inductions. Accordingly, just because the conclusions of inductions are neither justified a priori nor a posteriori, it does not follow that they are not justified.

1.3.2 The Second Premise

In this subsection, I criticize the second premise of Hume's argument against inductions that the uniformity principle is required for inductions to be justified, i.e., "all inferences from experiences suppose, as their foundation, that the future will resemble the past" (Hume, 1993, p. 24).

What Hume has in mind with the second premise can be illustrated by the following two arguments:

(I)
Observed gold bars are yellow.
∴ Unobserved gold bars are yellow.

(U)
Observed gold bars are yellow.
Nature is uniform.
∴ Unobserved gold bars are yellow.

According to Hume, (U) is intuitive while (I) “is not intuitive” (Hume, 1993, p. 22). In other words, (U) is justified while (I) is not.

In my view, it is controversial whether Hume is right on this count. He claims that (I) is not justified, but inductivists would claim that (I) is justified. On the inductivist account, (I) is justified even without the uniformity principle, i.e., the premise alone justifies the conclusion, and the premise does not need the uniformity principle to justify the conclusion. It is under dispute between Hume and inductivists whether (I) is justified. Under such circumstances, it begs the question against inductivists for Hume to say that (I) is not justified.

Hume would reply that (I) is not justified because its conclusion is neither justified a priori nor a posteriori. In Sect. 1.3.1 above, however, I objected that on the undemanding conception of justification, the conclusions of inductions might be justified although they might be false, that the faculty of reason and experience might jointly justify the conclusions of inductions, and thus that just because the conclusions of inductions are neither justified a priori nor a posteriori, it does not follow that they are not justified. It is merely an assumption that (I) is not justified.

Consider also that (I) is an inductive argument while (U) is a deductive argument. Thus, to say that (I) is not justified while (U) is justified is to say that an induction is not justified while a deduction is justified. It is circular to say so in the course of establishing the conclusion that inductions are not justifiable.

However, some philosophers (Lee, 2022) agree with Hume that (I) is not justified, and that the uniformity principle is required for (I) to be justified, and then attempt to justify the uniformity principle. In my view, to agree with the inductive skeptic on this count is to unjustifiably concede too much to him. Once you make the concession to him, you are doomed to be defeated by him, i.e., you will end up embracing his conclusion that inductions are not justifiable. To put it conversely, as long as you do not make the concession to him, you are entitled to reject his conclusion that inductions are not justifiable.

Let me raise a more serious objection to the second premise that the uniformity principle is required for inductions to be justified. Consider the relationship between particular beliefs—beliefs about particular objects—and a general belief—a belief about a set of objects. Are particular beliefs required for a general belief to be justified? Or is a general belief required for particular beliefs to be justified? Compare the following two arguments, (A) and (B):

(A)
This gold bar is yellow.
That gold bar is yellow.

...
∴ All gold bars are yellow.

(B)
All gold bars are yellow.
∴ This gold bar is yellow.
∴ That gold bar is yellow.
...

Whish of these two arguments reflects our cognitive lives? (A) accords with how we arrive at the general belief that all gold bars are yellow. We first observe some gold bars. As a result, we believe that all gold bars are yellow. The more gold bars we observe, the more we become convinced that all gold bars are yellow. Accordingly, it is accurate to say that particular beliefs are required for the general belief to be justified.

By contrast, (B) does not accord with how we arrive at a particular belief. We believe that this gold bar is yellow as a result of observing it, and the justification for the belief comes from experience. It is inaccurate to say that we are born with the general belief that all gold bars are yellow, that the general belief is justified independently of experience, and that the general belief justifies each particular belief that a gold bar is yellow. Rather, we form the general belief as a result of forming particular beliefs about particular gold bars. Consequently, it is inaccurate to say that the general belief is required for particular beliefs to be justified.

Just as (A) accords with, while (B) does not, our cognitive lives, so (1) comports well with, while (2) does not, our cognitive lives:

(1)
Since observed gold bars are yellow, unobserved gold bars are also yellow.
Since observed humans are mortal, unobserved humans are also mortal.
Since observed stones fall down, unobserved stones also fall down.
...
∴ Nature is uniform.

(2)
Nature is uniform.
∴ Since observed gold bars are yellow, unobserved gold bars are also yellow.
∴ Since observed humans are mortal, unobserved humans are also mortal.
∴ Since observed stones fall down, unobserved stones also fall down.
...

(1) comports well with how we arrive at the uniformity principle. We first form inductions. As a result, we form the general belief that nature is uniform. The more inductions we form, the more we become convinced that nature is uniform. Accordingly, it is accurate to say that inductions are required for the uniformity principle to be justified.

In contrast, (2) does not comport well with how we arrive at a particular induction. We believe that since observed gold bars are yellow, unobserved gold bars are yellow, as a result of observing some gold bars and inferring that unobserved gold bars are yellow. It is inaccurate to say that we are born with the uniformity principle, that the uniformity principle is justified independently of experience, and that the uniformity principle justifies inductions. Rather, we form the uniformity principle as a result of forming inductions (Park, 2023, pp. 478–479). Consequently, it is inaccurate to say that the uniformity principle is required for inductions to be justified.

In my view, (B) is merely a product of confusing a means and an end. Particular beliefs are the means, and the general belief is the end, i.e., particular beliefs are required for the general belief be justified. However, (B) takes the general belief to be the means, and particular beliefs to be the end, i.e., (B) assumes that the general belief is required for particular beliefs to be justified. The inevitable consequence of confusing the means and the end is skepticism that both particular belief and the general belief are not justifiable.

Similarly, (2) is merely a product of confusing a means and an end. Inductions are the means, and the uniformity principle is the end, i.e., inductions are required for the uniformity principle to be justified. However, (2) takes the uniformity principle to be the means, and inductions to be the end, i.e., (2) assumes that the uniformity principle is required for inductions to be justified. The inevitable consequence of confusing the means and the end is skepticism about both inductions and the uniformity principle.

To use an analogy, the suggestion that the uniformity principle is required for inductions to be justified is similar to the suggestion that being on the roof is required for using a ladder. In reality, to use a ladder is a means and to be on the roof is the end rather than the other way round. However, if you think erroneously that being on the roof is required for using the ladder, you confound the means and the end. The inevitable consequence of confusing the means and the end is that you can neither use the ladder nor be on the roof.

This analogy is intended to clarify rather than to justify my objection to the second premise of Hume's argument against inductions that it is to confuse a means and an end to say that the uniformity principle is required for inductions to be justified. Therefore, to attack the analogy is not to undermine my objection to the second premise but rather to undermine my method to clarify my objection to the second premise. Bear in mind that even if the analogy were destroyed, the objection might stand.

Confusing a means and an end can be found not only in Hume's argument against inductions but also in other philosophers' arguments, as we will see in the subsequent chapters of this book. For example, it can be found in Byeong D. Lee's (2022) coherentist argument for the uniformity principle (Chap. 2), in John Norton's (2003) localist argument for inductions (Chap. 3), in Bas van Fraassen's (1989, p. 143) argument against IBE (Chap. 4), Ruth Weintraub's (2013, p. 213) argument against IBE (Chap. 4), and Christophe de Ray's (2022) argument against IBE (Chap. 4).

1.3.3 The Third Premise

In this subsection, I criticize the third premise of Hume's argument against inductions that the uniformity principle is neither justified a priori nor a posteriori.

As noted in Sect. 1.2 above, Hume considers the possibility that the faculty of reason might justify the uniformity principle, and then rejects it on the ground that negating it does not give rise to a contradiction, i.e., ultimately on the grounds that it might be false. He also considers the possibility that experience might justify it, and then rejects it on the grounds that it is circular to appeal to experience to justify it. Surprisingly, he does not consider the possibility that the faculty of reason and experience might jointly justify it.

Why is it more plausible to say that the faculty of reason and experience jointly justify the uniformity principle than to say that they separately justify the uniformity principle? We form the uniformity principle on the basis of inductions, as noted in Sect. 1.3.2 above. Inductions are joint products of the faculty of reason and experience, as noted in Sect. 1.3.1 above. It follows that the uniformity principle is also a joint product of the faculty of reason and experience. Thus, the uniformity principle is more likely to be justified jointly justified by the faculty of reason and by experience than by the faculty of reason alone or by experience alone.

Hume claims that the uniformity principle is required for inductions to be justified. Under his framework, it is circular to provide inductions to justify the uniformity principle, and the problem of induction arises. There is an alternative framework according to which the faculty of reason and experience jointly justify inductions, and inductions justify the uniformity principle. Under this alternative framework, it is not circular to provide inductions to justify the uniformity principle, and the problem of induction does not arise.

1.3.4 The Inference

In Sect. 1.2 above, I summarized Hume's argument against inductions in the following standard form:

Inductions are neither justified a priori nor a posteriori.
The uniformity principle is required for inductions to be justified.
The uniformity principle is neither justified a priori nor a posteriori.
∴ Inductions are not justifiable.

In this subsection, I argue that the inference from the premises to the conclusion conflicts with what the conclusion asserts. I also anticipate a possible response from Hume and then construct a dilemma for him.

As noted in Sects. 1.2 and 1.3.1 above, Hume explores the possibility that the faculty of reason might justify the conclusions of inductions, and then rejects it on the grounds that the negations of the conclusions of inductions do not involve contradictions. His first attempt to find a justifier for the conclusions of inductions fails. He then explores the possibility that experience might justify the conclusions of inductions, and then rejects it on the grounds that cause and effect are not necessarily connected with each other. His second attempt to find a justifier for the conclusions of inductions fails. He then concludes that inductions are not justified.

What are we to make of this inference? The conclusion implies that future attempts to find a justifier for the conclusions of inductions will also fail. In a nutshell, he runs the following inductive argument to show that inductions are not justifiable:

(H1)
His two past attempts to find a justifier for inductions failed.
∴ Future attempts to find a justifier for inductions will also fail.

(H1) is similar to the inductive argument that since two observed gold bars are yellow, unobserved gold bars are also yellow. It is self-defeating for Hume to appeal to (H1)—an induction—to show that inductions are not justifiable.

A similar story can be given about the inference from the third premise to the conclusion. As noted in Sect. 1.3.3 above, Hume explores the possibility that the faculty of reason might justify the uniformity principle, and then rejects it on the grounds that the negation of the uniformity principle does not involve a contradiction. His first attempt to find a justifier for the uniformity principle fails. He then explores the possibility that experience might justify the uniformity principle, and then rejects it on the grounds that it is circular to appeal to experience to justify the uniformity principle. His second attempt to find a justifier for the conclusions of inductions fails. He then concludes that the uniformity principle is not justified.

What are we to make of this inference? The conclusion implies that future attempts to find a justifier for the uniformity principle will also fail. In a nutshell, he runs the following inductive argument to show that the uniformity principle is not justified:

(H2)
His two past attempts to find a justifier for the uniformity principle failed.
∴ Future attempts to find a justifier for the uniformity principle will also fail.

(H2) is similar to the inductive argument that since two observed gold bars are yellow, unobserved gold bars are also yellow. It is self-defeating for Hume to appeal to (H2)—an induction—to establish that inductions are not justifiable.

How would Hume reply to the foregoing objection to his argument for the problem of induction? He might argue that (H1) is a second-order induction while the inductions that (H1) is intended to show are not justified are first-order inductions, i.e., the term "inductions" in (H1) refers to first-order inductions, but not to (H1). First-order inductions concern objects and events in the world, while (H1) concerns first-order inductions in our minds. It follows that (H1) is separate from first-order inductions, and its content is not entailed by the contents of first-order inductions. Consequently, it is not self-defeating to provide (H1) to show that first-order inductions are not justifiable.

This possible response from Hume, although reasonable, is similar to the standard Humean objection to the following second-order arguments for first-order inductions and for the uniformity principle:

(SF)
Inductions have worked in the past.
∴ They will also work in the future.

(SU)
The uniformity principle was supported by experience in the past.
∴ The uniformity principle will also be supported by experience in the future.

(SF) is a second-order induction for first-order inductions, and (SU) is a second-order induction for the uniformity principle. Humean skeptics object that (SF) and (SU) are inductions, and that it is circular to provide them to show that inductions are justified.

But is it really circular for inductivists to appeal to (SF) and (SU) to show that inductions are justified? For example, (SF) is a second-order induction, while the inductions that (SF) is intended to show are justified are first-order inductions, i.e., the term "inductions" in (SF) refers to first-order inductions, but not to (SF). It follows that (SF) is separate from first-order inductions, and that its content is not entailed by the contents of first-order inductions. Accordingly, it is not circular to provide (SF) to show that first-order inductions are justified.

This conclusion would not be convincing to staunch inductive skeptics. They would maintain that (SF) and first-order inductions are all inductions, and thus that it is circular to provide (SF) to show that first-order inductions are justified. I invite staunch inductive skeptics to consider that their position is vulnerable to the following two objections.

First, if it is circular for inductivists to provide (SF) to show that inductions are justified, it is self-defeating for Hume to provide (H1) to show that inductions are not justifiable, given that both (SF) and (H1) are all inductions. It is arbitrary for staunch inductive skeptics to say that it is circular for inductivists to provide (SF) to show that inductions are justified, but that it is not self-defeating for Hume to provide (H1) to show that inductions are not justifiable.

Second, if it were circular to provide (SF) to show that first-order inductions are justified because they are the same kind of inferences, it would also be circular to provide p to show that q is justified, whatever p and q might be, because they are the same kind of things, viz., propositions, and it would also be circular to appeal to perceptions (nondoxastic mental states) to show that perceptual beliefs about observed objects are justified because perceptions and perceptual beliefs are the same kind of things, viz., mental states.

To go further, unless the justifier and the justifiee (that which is justified) were completely different kinds of things, it would be circular to provide the justifier to justify the justifiee. If the justifier and the justifiee are both existent things, the justifier cannot justify the justifiee because they are the same kind of things, viz., existent things. It follows that either the justifier or the justifiee should not exist in order for the justifier to justify the justifiee. However, it is absurd that a non-existent thing can justify an existent thing, or that an existent thing can justify a non-existent thing. This absurd consequence follows from staunch inductive skeptics' position that it is circular to provide (SF) to show that first-order inductions are justified.

In any event, I construct the following dilemma for Hume. He can either embrace or resist my foregoing defense that it is not circular to provide (SF) to show that first-order inductions are justified. On the one hand, if he embraces it, he cannot say that it is circular for inductivists to provide (SF) to show that inductions are justified. Nor can he say that inductions are not justified. On the other hand, if he resists it, he can say that it is circular for inductivists to provide (SF) to show that inductions are justified, but he faces my objection that it is self-defeating for him to provide (H1) and (H2) to establish that inductions are not justifiable. In short, either it is not circular inductivists to provide (SF) to show that inductions are justified, or Hume's argument against inductions is self-defeating.

1.3.5 Why Construct an Argument for or Against Inductions?

In this subsection, I expose fundamental problems with constructing an argument to show that inductions are justifiable or unjustifiable.

Imagine that skeptics construct the following argument against all of our beliefs. Any of our beliefs, whatever it is, is justified or not, depending on whether the mind is reliable or not, i.e., if the mind were reliable, it would be justified, and if the mind were not reliable, it would not be justified. However, we cannot show that the mind is reliable. After all, the use of the mind is required to show that the mind is reliable, and it is circular to use the mind to show that the mind is reliable. In consequence, none of our beliefs is justifiable.

The foregoing argument against all of our beliefs is similar to Hume's argument against inductions in that they make use of the same idea that it is circular to use X to show that X is justified. The argument against all of our beliefs asserts that it is circular to use the mind to show that the mind is reliable. Hume's argument against inductions asserts that it is circular to use an induction to show that inductions are justified. It appears, therefore, that no one can show that the mind is reliable, or that inductions are justifiable.

Does it follow that we should despair and give up all of our beliefs? A negative answer to this question can be defended as follows. The argument against all of our beliefs and Hume's argument against inductions are both self-defeating. The use of the mind is required to draw the conclusion that the mind is not reliable; Hume uses inductions, (H1) and (H2), to draw the conclusion that inductions are not justifiable, as noted in Sect. 1.3.2 above. Therefore, just as no one can show that the mind is not reliable, so no one can show that inductions are not justifiable.

Some readers might not yet be convinced of my contention that no one can show that inductions are not justifiable. I ask them to investigate whether the following argument is deductive or inductive:

Inductions are neither justified a priori nor a posteriori.
The uniformity principle is required for inductions to be justified.
The uniformity principle is neither justified a priori nor a posteriori.
∴ Inductions are not justifiable.

This argument is not deductive but inductive. The mere possibility that the faculty of reason and experience might jointly justify inductions shows that even if the premises are true, the conclusion might be false. It is self-defeating for Hume to advance such an argument to show that inductions are not justifiable.

Hume's defenders might now object that I have misunderstood Hume's argument against inductions, and that the standard form above misrepresents Hume's argument against inductions. They might then present their own interpretations of Hume's argument against inductions, and then conclude that none of my objections to it in this chapter succeeds.

In response, I construct the following dilemma against *any* argument whose conclusion is that inductions are not justifiable. The argument is either deductive or inductive. On the one hand, if it is deductive, it begs the question against inductivists. To state it differently, its premises assume the very point that needs to be established, viz., its conclusion that inductions are not justifiable. Hence, new arguments need to be constructed to justify its premises. If the new arguments for its premises are deductive, the problem of infinite regress is in the offing. On the other hand, if the argument is inductive, it is self-defeating to advance it to show that inductions are not justifiable. Therefore, it is in principle impossible to undermine inductions, i.e., to construct any argument that would show that inductions are not justifiable.

This conclusion applies to all the interpretations of Hume's argument against inductions, whatever they might be. Is Hume's argument deductive or inductive? I ask this simple question to my future opponents who aim to show that I misinterpreted Hume's argument against inductions, and thus that all of my objections to Hume's argument against inductions in this chapter misfire.

To the dismay of inductivists, however, a similar dilemma can also be constructed against *any* argument whose conclusion is that inductions are justified. The argument is either deductive or inductive. On the one hand, if it is deductive, it begs the question against inductive skeptics. To state it otherwise, its premises assume the very point that needs to be established, viz., its conclusion that inductions are justified. Hence, new arguments need to be constructed to justify its premises. If the new arguments for its premises are deductive, the problem of infinite regress is in the offing. On the other hand, if the argument is inductive, it is circular to provide it to justify inductions. Therefore, it is in principle impossible to justify inductions, i.e., to construct any argument that would show that inductions are justified.

This conclusion applies to all the arguments that philosophers have constructed to justify inductions in the literature ever since Hume advanced the problem of induction, and to all arguments that philosophers will construct to justify inductions. Is your argument deductive or inductive? I ask this simple question to those who have constructed or will construct an argument to show that inductions are justified. In my view, to attempt to justify inductions is to undertake a task that it is in principle impossible to be accomplished. This conviction will be reinforced in Chaps. 2 and 3 where I criticize Lee's (2022) coherentist attempt and John Norton's (2003) localist attempt to justify inductions, respectively.

Even if no one can show that inductions are justified, they might nonetheless be justified. How does this follow? There are internalist and externalist theories of justification in epistemology. Internalism affirms, while externalism denies, that a justifier for p should be accessible to you in order for you to be justified in believing that p (Goldman & Beddor, 2021). Externalists would say that inductions are justified or unjustified, not depending on whether we have identified justifiers for inductions or not, i.e., not depending on whether we have succeeded in showing or failed to show that they are cogent, but rather depending on whether they are cogent or not, i.e., depending on whether the premises are true, and whether the conclusions probably follow from the premises. If the two conditions are met, they are justified even if we do not know that the two conditions are met, and even if we cannot recognize justifiers for inductions.

This subsection can be summed up as the following simple question to those who aim to solve the problem of induction, i.e., to those who aim to justify inductions. Why do you think that inductions need to be justified? As noted in Sects. 1.3.1 and 1.3.2, it is wrong to answer that inductions need to be justified because they are neither justified a priori nor a posteriori, or because the uniformity principle is neither justified a priori nor a posteriori.

1.3.6 The Backfire

The conclusion of Hume's argument against inductions holds that inductions are not justifiable. Let me elucidate its disastrous consequence for his positive account of why we predict future events.

We regularly predict future events. For example, after observing that overeating and obesity were conjoined in the past, we predict that they will be conjoined in the future. Why do we predict that they will be conjoined in the future? Hume considers the following two rival hypotheses. One is that we are determined by the faculty of reason to make the prediction. The other is that we are determined by the habit to make the prediction. He claims that we are not determined by the faculty of reason but rather "we are determined by custom alone to expect the one from the appearance of the other" (Hume, 1993, p. 28). Note that he uses IBE to arrive at his view that we make future predictions by habit.

Hume's argument against inductions militates against his own use of IBE. The premise of his IBE is that we make predictions on the basis of past experience, and the conclusion of his IBE is that we have the habit of making predictions on the basis of past experience. This inductive inference is neither justified a priori nor a posteriori. It is not justified a priori because no contradiction is involved in the negation of the conclusion. Neither is it justified a posteriori because there is no necessary connection between making predictions and forming the habit. The inference from the premise to the conclusion requires the uniformity principle which implies that whenever we make predictions, we have the habit. However, the uniformity principle is neither justified a priori nor a posteriori. Therefore, Hume's conclusion that we have the habit is "entirely arbitrary" (Hume, 1993, p. 18). In short, he is hoisted on his own petard.

What can we learn? Skepticism has several epistemic and pragmatic disadvantages (Park, 2019, pp. 96–97). One of them is that the less you believe, the less you can explain. Hume is skeptical about inductions. As a result, he cannot explain anything, and he cannot be relieved of the puzzlement over why certain events occur around us. For example, why do we predict the future? Why do we become fat? Why do eclipses occur? He cannot answer these questions because answering them requires the use of IBE. This disadvantage of his position supports my previous slogan: "Believe little. Explain little" (Park, 2015, p. 28).

1.4 The Problem of Counter-Induction

I anticipate that my critics will raise brilliant objections to inductions, and then request that I defend inductions against those objections. For example, they might advance a counter-induction (Salmon, 1957, p. 46) that since observed gold bars are yellow, unobserved gold bars are not yellow, and then request me to defend the induction that since observed gold bars are yellow, unobserved gold bars are also yellow.

First of all, this request to justify inductions is unfair in light of the dilemma that I constructed in Sect. 1.3.5 above. The conclusion of the dilemma is that it is in principle impossible to construct an argument that would show that inductions are justified. As far as I am concerned, to request that I justify inductions is to request that I achieve something that it is in principle impossible to achieve. My critics are asked to refute the dilemma before they request that I justify inductions.

Regarding the counter-induction, I develop a philosophical problem against it that I call "the problem of counter-induction" according to which counter-inductions are not justifiable. My argument for the problem of counter-induction goes as follows. Counter-inductions are not justifiable a priori because the negations of their conclusions do not involve contradictions. Counter-inductions are not justifiable a posteriori either because there is no necessary connection between cause and counter-effect. For example, overeating is not necessarily connected with remaining slim. Hence, experience does not justify the conclusions of counter-inductions. Counter-inductions require the disuniformity principle that nature is disuniform (Park, 2017, 2018, 2022, pp. 76–79). The disuniformity principle is not justified a priori because its negation does not involve a contradiction. The disuniformity principle is not justified a posteriori either because it is circular to appeal to experience to justify the disuniformity principle. Therefore, counter-inductions are not justifiable.

The foregoing argument against counter-inductions can be summed up in the following standard form:

Counter-inductions are neither justified a priori nor a posteriori.
The disuniformity principle is required for counter-inductions.
The disuniformity principle is neither justified a priori nor a posteriori.
∴ Counter-inductions are not justifiable.

The argument against counter-inductions mirrors Hume's argument against inductions. Two noteworthy consequences follow from the similarity between them. First, it is arbitrary to say that inductivists have the burden of critically responding to the problem of induction, but that counter-inductivists do not have the burden of critically responding to the problem of counter-induction. Second, counter-inductivists might come up with brilliant ideas to defend counter-inductions vis-à-vis the argument against counter-inductions. Whatever ideas they might come up with, inductivists would make use of them to defend inductions vis-à-vis Hume's argument against inductions.

Admittedly, my foregoing critical response to the counter-induction does not amount to the justification of inductions. It merely amounts to a tu quoque. However, my tu quoque against the counter-induction is consistent with my earlier statement in Sect. 1.3.5 above that it is in principle impossible to construct an argument that would show that inductions are justified, and with my earlier statement in Sect. 1.1 that this chapter does not aim to show that inductions are justified but rather aims to criticize Hume's argument against inductions. Those who aim to strengthen his argument against inductions with the use of counter-induction have the burden of responding to the problem of counter-induction sketched in this section.

Incidentally, I do not believe that we should always be inductivists. Nor do I believe that we should always be counter-inductivists. Rather, I believe that we should be inductivists in some cases and counter-inductivists in other cases. An interesting question is under what conditions we should choose inductions over counter-inductions and vice versa. I answer this question in Chap. 7.

1.5 Conclusion

My key objections to Hume's argument against inductions are as follows. The faculty of reason and experience might jointly justify inductions. It is self-defeating for Hume to use inductions to establish that inductions are not justifiable. Inductions are required for the uniformity principle to be justified rather than the other way around. It is in principle impossible to construct an argument that would show that inductions are justified or unjustified. I conclude this chapter with the following two slogans: "Why attempt to justify inductions?" "The problem of induction is a product of confusing a means and an end."

References

de Ray, C. (2022). A new epistemological case for theism. *Religious Studies, 58*(2), 379–400.
Goldman, A., & Beddor, B. (2021). Reliabilist epistemology. In E. N. Zalta (Ed.), *The Stanford encyclopedia of philosophy*. Retrieved from https://plato.stanford.edu/archives/sum2021/entries/reliabilism/
Hume, D. (1993). *An enquiry concerning human understanding*. Hackett Publishing.
Lee, B. (2022). A coherentist justification of induction. *Erkenntnis, 87*(1), 35–52.
Norton, J. (2003). A material theory of induction. *Philosophy of Science, 70*(4), 647–670.
Park, S. (2015). Explanatory failures of relative realism. *Epistemologia, 38*(1), 16–28.
Park, S. (2017). The uniformity principle vs. the disuniformity principle. *Acta Analytica, 32*(2), 213–222.
Park, S. (2018). The anti-induction for scientific realism. *Grazer Philosophische Studien, 95*(3), 329–342.
Park, S. (2019). The disastrous implications of the 'English' view of rationality in a social world. *Social Epistemology, 33*(1), 88–99.
Park, S. (2022). *Embracing scientific realism*. Springer.
Park, S. (2023). Inference to the best explanation, naturalism, and theism. *Implicit Religion, 24*, 472–485.
Salmon, W. (1957). Should we attempt to justify induction? *Philosophical Studies, 8*(3), 33–48.
van Fraassen, B. (1989). *Laws and symmetry*. Oxford University Press.
Weintraub, R. (2013). Induction and inference to the best explanation. *Philosophical Studies, 166*(1), 203–216.

Chapter 2
Can Coherentism Solve the Problem of Induction?

Abstract I raise the following objections to Lee's coherentist justification of induction. Coherentism is not an adequate theory of epistemic justification. It is circular for Lee to use inductions to justify the uniformity principle. It is one thing for the uniformity principle to be justified in the coherentist sense; it is quite another for the uniformity principle to be likely to be true. It begs the question against Hume for Lee to take the uniformity principle to be a default belief. Like Hume's argument against inductions, Lee's coherentist justification of induction confounds a means and an end.

Keywords Coherentism · Defeater · Induction · Inference to the best competitor · Uniformity principle

2.1 Introduction

Ever since David Hume (1993, pp. 15–25) raised the problem of induction, many philosophers have attempted to solve it, i.e., to justify inductions. This chapter focuses on Byeong D. Lee's (2022) attempt to justify inductions. He develops a new variant of the coherentist theory of epistemic justification (referred to hereafter as "coherentism") and then applies it to the uniformity principle with the intention of showing that the uniformity principle is justified, and hence that inductions are justified. This strategy for solving the problem of induction is original, and it deserves critical scrutiny. This chapter aims to show that it does not succeed in showing that inductions are justified.

This chapter is structured as follows. In Sect. 2.2, I sketch what a solution to the problem of induction would look like. In Sect. 2.3, I unpack Lee's attempt to solve the problem of induction, i.e., his argument for the uniformity principle. In the remaining sections, I raise multiple objections to his argument. My main objections are as follows. Coherentism is not an adequate theory of epistemic justification. It is circular for Lee to use inductions to justify the uniformity principle. It is one thing for the uniformity principle to be justified in the coherentist sense; it is quite another

S. Park, *Induction, Science, and Morality*, Synthese Library 520,
https://doi.org/10.1007/978-3-032-00032-3_2

for the uniformity principle to be likely to be true. It begs the question against Hume to take the uniformity principle to be a default belief. Like Hume's argument against inductions, Lee's coherentist justification of induction confounds a means and an end.

Readers will see how coherentists might attempt to show that inductions are justified and how Hume would respond to their attempt. As a result, they will gain a better understanding of the relationship between inductions and coherentism.

2.2 What a Solution to the Problem of Induction Look like

The problem of induction holds that inductions are not justifiable. In this section, I summarize Hume's argument against inductions and then sketch what a solution to the problem of induction would look like.

David Hume (1993, pp. 15–25) constructs an argument to show that inductions are not justified, which I analyze in detail in Sect. 1.2 of Chap. 1. To put it briefly, the conclusions of inductions are not justified a priori because their negations do not involve contradictions (Hume, 1993, p. 15). Nor are they justified a posteriori because cause and effect are not necessarily connected with each other (Hume, 1993, p. 21). The uniformity principle is required for inductions to be justified (Hume, 1993, p. 22). The uniformity principle is not justifiable a priori because it is not a contradiction to say that nature is not uniform (Hume, 1993, p. 22). Nor is it justified a posteriori because it is circular to appeal to experience to justify the uniformity principle (Hume, 1993, p. 23). Therefore, inductions are not justifiable.

What would a solution to the problem of induction look like? Given that the conclusion of Hume's argument against inductions states, as just noted, that inductions are not justifiable, a solution to the problem of induction would consist in showing that inductions are justified, i.e., in showing that inductions are cogent. Inductions are cogent when the following two conditions are met. First, their premises are true. Second, their conclusions probably follow from their premises. The problem of induction is solved, once existing inductions, such as the induction that since observed pine trees are green, unobserved pine trees are also green, are shown to have met the two conditions.

A solution to the problem of induction does not consist in making uncogent arguments cogent. Cogent arguments can become uncogent, and uncogent arguments can become cogent, depending on whether new premises are added to them and/or existing premises are removed from them. Hume neither adds new premises to existing inductions nor removes existing premises from existing inductions. Therefore, a solution to the problem of induction does not consist in making changes to existing inductions.

Existing inductions are cogent or uncogent independently of Hume's argument against them and inductivists' argument for them. It is wrong to think that existing inductions were cogent, but they have become uncongent due to Hume's argument against them. It is also wrong to think that existing inductions were uncogent, but they have become cogent thanks to inductivists' argument for them. Existing inductions are cogent or uncogent, not depending on whether philosophers have strong arguments for or against them, but rather depending on whether they have met the aforementioned two conditions. Therefore, even if we grant for the sake of argument that Hume's argument against inductions is strong, we can still say that existing inductions might be cogent, i.e., that they might be justified.

2.3 Lee's Argument for the Uniformity Principle

In this section, I sketch how Lee utilizes coherentism to show that inductions are justified, i.e., that they are cogent.

Lee (2022) develops a new form of coherentism, drawing on insights from Wilfrid Sellars (1963), Keith Lehrer (2000), and Robert Brandom (1994). It holds that we have the epistemic goal of obtaining "a world picture with a maximum of explanatory coherence" (Lee, 2022, p. 38). We are "justified in accepting that *p* just in case we can answer all objections raised against it in our social practice of justification" (Lee, 2022, p. 39), or just in case p can "beat all competing hypotheses" (Lee, 2022, p. 43). To justify p "requires that some claims be treated as having default justification unless any challenger provides some positive reasons to doubt them" (Lee, 2022, p. 40).

Lee applies his version of coherentism to the uniformity principle to show that it is justified, and hence that inductions are justified. His argument for the uniformity principle (UP) consists of the following two premises and conclusion. (1) We have "the epistemic goal of explaining and predicting the world" (Lee, 2022, p. 41). (2) We "have no way but to rely on UP in order to predict the future in a rational way" (Lee, 2022, p. 41). This premise indicates that we can answer all objections to the uniformity principle, or that the uniformity principle can beat all competitors, in our social practice of justification. (3) Therefore, "it is reasonable for us to accept UP" (Lee, 2022, p. 41). He takes this argument to be "a coherentist solution to Hume's problem of induction" (Lee, 2022, p. 37). He also claims that his "coherentist justification of induction provides a new viable solution to the problem of induction" (Lee, 2022, p. 51).

In my view, showing that the uniformity principle is justified would indeed constitute a solution to the problem of induction. After all, if the uniformity principle were justified, it could be used, along with the premises of inductions, to justify the conclusions of inductions. As a result, the conclusions of inductions would be shown to be likely to be true,[1] and inductions would be shown to be cogent, contrary to what Hume thinks. As such, the issue comes down to whether Lee has succeeded in showing that the uniformity principle is justified.

In what follows, I criticize Lee's foregoing strategy to solve the problem of induction, i.e., to justify inductions.

---

[1]As noted in Sect. 1.3.1 of Chap. 1, Alvin Goldman and Bob Beddor (2021) claim that a justified belief is likely to be true.

2.4 The Dilemma

In this section, I construct a dilemma against Lee's attempt to justify inductions.

In Sect. 1.3.5 of Chap. 1, I constructed the following dilemma against any argument intended to show that inductions are justified. The argument is either deductive or inductive. On the one hand, if it is deductive, its premises assume its conclusion, and hence they need to be justified. On the other hand, if it is inductive, it is circular to provide it to show that inductions are justified. Therefore, it is in principle impossible to construct an argument that would show that inductions are justified.

This dilemma applies to Lee's coherentist argument for the uniformity principle summarized in Sect. 2.3 above. His argument is either deductive or inductive. On the one hand if it is deductive, its premises need to be justified. On the other hand, if it is inductive, it is circular to provide it to show that the uniformity principle is justified. After all, the uniformity principle is invoked to justify inductions. To use an induction to justify the uniformity principle amounts to using an induction to justify inductions. Therefore, it is in principle impossible for Lee's coherentist argument for the uniformity principle to get off the ground.

This dilemma entitles us to predict that engaging in the details of Lee's coherentist argument for the uniformity principle will expose specific problems with it. Readers will confirm this prediction in the subsequent sections.

2.5 The Epistemic Goal?

In this section, I criticize the teleological aspect of Lee's argument for the uniformity principle.

The first premise of Lee's argument for the uniformity principle states that we have "the epistemic goal of explaining and predicting the world" (Lee, 2022, p. 41). Relatedly, Lee states that "insofar as it is reasonable for us to pursue our epistemic goal of predicting the future, it is reasonable for us to accept UP" (Lee, 2022, p. 41). To put it bluntly, since we have the epistemic goal of predicting the future, we can justifiably accept the uniformity principle.

How would Hume respond to this argument for the uniformity principle? He would point out that Lee should first attempt to refute his argument against the uniformity principle before constructing an argument for it. Hume's argument against it is that it is not justified a priori given that it is not a contradiction to say that nature is disuniform. Nor is it justified a posteriori given that it is circular to rely on experience to justify the uniformity principle.

In addition, Hume might present the following counterexample against Lee's argument for the uniformity principle. Prosecutors have the epistemic goal of believing that defendants are guilty. It does not follow, however, that they are justified in believing that defendants are guilty. Whether a belief is justified depends not on what believers' epistemic goal is, but on whether it meets a certain condition under which it is likely to be true, or whether it has a certain property that makes it likely to be true. Believers' epistemic goal is neither the condition nor the property.

Hume might add that an epistemic goal, if anything, makes cognitive agents biased and their beliefs less justified. Prosecutors have the epistemic goal of defending their position that defendants are guilty, and thus they might be tempted to jump to their favored conclusion, in the absence of sufficient evidence, that defendants committed crimes. Similarly, we have the epistemic goal of predicting the future, and thus we might be tempted to jump to our favored conclusion, in the absence of sufficient evidence, that a certain event will occur in the future. In short, an epistemic goal may have an adverse effect on the justificatory status of beliefs. So might Hume argue against Lee.

Lee claims that "justification is a goal-dependent concept" (Lee, 2022, p. 38). On his account, p is justified or unjustified, depending on what our epistemic goal is. I call this view of justification "the teleological conception." It contrasts with what I call "the mechanical conception." The teleological conception invokes a goal, while the mechanical conception invokes a cause, to explicate the concept of justification. Alvin Goldman (1979) advocates the mechanical conception, when he proposes that a belief is justified if and only if it is caused by a reliable cognitive process, a process that produces a high frequency of true beliefs.

Philosophers of science similarly distinguish between teleological and mechanical explanations. The former invoke goals, while the latter invoke causes, to explain phenomena. For example, it is a teleological explanation that we have eyebrows so that our eyes are protected from raindrops. It is a mechanical explanation that we have eyebrows because our distant ancestors acquired eyebrows as a result of variations, eyebrows were adaptive traits, and they were passed on to next generations. Ancient scientists, such as Aristotle, regarded both kinds of explanations as legitimate. However, modern scientists, such as Darwin, take only mechanical explanations as legitimate. Teleological explanations have no place in modern science.

Why are teleological explanations illegitimate? A goal concerns a future event. Lee claims, for example, that we have the epistemic goal "to gradually improve our conceptual framework, so as to achieve a world picture with a maximum of explanatory coherence" (Lee, 2022, p. 38). The world picture with the maximum explanatory coherence does not exist now; it might or might not exist in the future. It is problematic to explain why a present belief is justified in terms of a future event that may or may not occur. In addition, Hume is skeptical about an inductive inference from the past to the future. Consequently, he would be skeptical about teleological explanations.

I predict that just as teleological explanations of phenomena are no longer considered to be legitimate in science, so the teleological conception of justification will no longer be considered to be legitimate in philosophy. Admittedly, Hume would take my prediction to be entirely arbitrary on the grounds that it is the conclusion of an induction. In Sect. 1.3 of Chap. 1, however, I raised six objections to his argument against inductions. In Sect. 1.3.1, for example, I objected that he overlooks the possibility that the faculty of reason and experience might jointly justify the conclusions of inductions. Therefore, he is not entitled to say that my prediction is not justified.

2.6 The Concept of Defeat

In this section, I criticize the concept of defeat, the key concept in the second premise of Lee's argument for the uniformity principle, arguing that coherentism is not an adequate theory of epistemic justification and cannot be used to justify the uniformity principle.

Lee wholeheartedly endorses Lehrer's concept of coherence, saying that "Lehrer (2000, esp. p. 170) offers us a very illuminating account of this elusive concept. On his proposal, we can understand coherentist justification in terms of 'answering all objections' or 'beating all competitors'" (Lee, 2022, p. 39). Lee suggests that the uniformity principle defeats all objections to it or beats all competitors in our social practice of justification, and thus that it is justified. This attempt to justify the uniformity principle is intriguing.

It is problematic for Lehrer to explicate the notion of justification in terms of the notion of beating all competitors. Suppose that two hypotheses, H_1 and H_2, compete with each other, and that H_1 beats H_2. It does not follow, however, that H_1 is justified. It only follows that H_1 is better justified than H_2. It is one thing that H_1 is better justified than H_2; it is another that H_1 is justified. After all, it might be that neither H_1 nor H_2 is justified, but that H_1 is slightly better justified than H_2. To use an analogy, suppose that Alice is taller than Diana. It does not follow that Alice is tall. It might be that both are short, and Alice is slightly taller than Diana (Wray, 2008, p. 323; Mizrahi, 2013, p. 401). It is a relative claim that Alice is taller than Diana, while it is an absolute claim that Alice is tall. The absolute claim does not follow from the relative claim. Analogously, it is a relative claim that H_1 beats H_2. In contrast, it is an absolute claim that H_1 is justified. The absolute claim does not follow from the relative claim. Therefore, just because H_1 beats H_2, it does not follow that H_1 is justified.

The same holds for the uniformity principle and its competitors, e.g., the disuniformity principle (Park, 2017, 2018) that nature is disuniform. Just because the uniformity principle beats its competitors, it does not follow that it is justified; it only follows that it is better justified than its competitors. Consequently, the inference from the premises to the conclusion of Lee's coherentist argument for the uniformity principle is fallacious. Specifically, it is a fallacious inference that since the uniformity principle beats its competitors, it is justified.

Let me raise a more fundamental objection to the notion of defeat. On what grounds can we say that H_1 defeats H_2? It would be circular for coherentists to say that H_1 defeats H_2 because the evidence for H_1 is stronger than the evidence for H_2, i.e., because H_1 is better justified than H_2 in light of our social practice of justification. After all, the notion of defeat is originally provided to explicate the notion of justification. To appeal to the notion of better justification to explicate the notion of defeat amounts to appealing to the notion of better justification to explicate the notion of better justification. I leave to coherentists the task of fleshing out the notion of defeat without appealing to the notion of better justification.

Relatedly, how could Lee show that the uniformity principle beats its competitors? It would be circular for Lee to suggest that the uniformity principle beats its competitors because the uniformity principle is better justified than are its competitors. As noted above, it is circular to invoke the notion of better justification to explicate the notion of defeat. It would also be circular for Lee to use an induction to show that the uniformity principle beats its competitors. After all, his coherentist argument for the uniformity principle is originally provided to justify inductions. To use an induction to show that the uniformity principle beats its competitors is tantamount to using an induction to justify inductions. I leave to Lee the task of showing that the uniformity principle beats its competitors without relying on an induction.

Coherentists might suggest that the uniformity principle is justified in the sense that it beats its competitors. On this suggestion, to say that the uniformity principle is justified is to say that it beats its competitors, but not that it is likely to be true. It follows, by this definition of justification, that the uniformity principle is justified.

What are we to make of this suggestion? We do not need to dispute it; we only need to be careful when we interpret coherentists' statement that the uniformity principle is justified. Their statement does not mean that the uniformity principle is likely to be true, but rather that it beats its competitors. It is one thing for it to beat its competitors; it is quite another for it to be likely to be true. We should not slide from the former to the latter.

Hume would say that a solution to the problem of induction does not consist in showing that the uniformity principle beats its competitors, but rather in showing that it is likely to be true. Once it is shown to be likely to be true, it can be used, along with the premises of inductions, to show that the conclusions of inductions are likely to be true, i.e., to show that inductions are cogent.

2.7 The Alternatives to the Uniformity Principle

2.7.1 *How Does Lee Know?*

In this subsection, I argue that Lee implicitly uses an induction to justify the second premise that we have no way to predict the future but to use the uniformity principle to predict the future, and that it is circular for him to do so.

How does Lee know that we have no way to predict the future but to use the uniformity principle? He might answer that he has attempted in vain to find other ways. This possible answer implies that since his past attempts to find other ways have failed, his future attempts to find other ways will also fail. Note that the use of an induction is involved in this possible answer. Hence, this possible answer is not available to him, and the original question remains unanswered. He cannot answer that he knows a priori that we have no other ways but to use the uniformity principle to predict the future because no contradiction is involved in the suggestion that there are other ways to predict the future.

The same objection can be raised to Lee's contention that "we have no positive reason to think that this principle is false" (Lee, 2022, p. 41). How does he know that we have no reason to think that the uniformity principle is false? He might answer that he has attempted in vain to find such reasons. This possible answer implies that since his past attempts to find such reasons have failed, his future attempts to find such reasons will also fail. Note that the use of an induction is involved in this possible answer. Hence, this possible answer is not available to him, and the original question remains unanswered. He cannot answer that he knows a priori that we have no reason to think that the uniformity principle is false because no contradiction is involved in the statement that we have reason to think that the uniformity principle is false.

2.7.2 There Are Other Ways

In this subsection, I argue that there are many other ways to predict the future, contrary to what the second premise of Lee's argument for the uniformity principle asserts.

There is an alternative to the uniformity principle, viz., the disuniformity principle (Park, 2017, 2018) that nature is disuniform. Inductions are to the uniformity principle what counter-inductions are to the disuniformity principle. The existence of the disuniformity principle opens the possibility that there might be many other alternatives between the two extremes of the uniformity and disuniformity principles. One of them holds, for example, that nature is uniform in some respects, but not in other respects. Each of the alternatives between the two extreme principles makes a slightly different claim about the respects in which nature is uniform or disuniform. The existence of such alternatives conflicts with Lee's second premise that we have no way to predict the future but to use the uniformity principle. Lee has the burden of showing, without using an induction, that the uniformity principle beats all the alternatives. If he uses an induction, the problem of circularity will arise.

Lee might object that it is irrational to use the disuniformity principle to predict the future. However, counter-inductivists would reply that it is irrational to use the uniformity principle to predict the future. Lee might now argue that the uniformity principle is a stronger competitor than the disuniformity principle, given that the uniformity principle is consistent, while the uniformity principle is inconsistent, with our social practice of justification. Therefore, the uniformity principle beats the disuniformity principle. In my view, however, it is circular for Lee to appeal to our social practice of justification to justify the uniformity principle. After all, our social practice of justification takes inductions for granted, and thus appealing to our social practice of justification amounts to appealing to inductions to justify inductions.

2.8 The Problem of Inference to the Best Competitor

In this section, I criticize the coherentist rule of inference that I call "inference to the best competitor" (IBC), thereby showing that it is futile to appeal to IBC to justify inductions.

The second premise of Lee's coherentist argument for the uniformity principle holds that we have no way to predict the future but to use the uniformity principle. The conclusion holds that the uniformity principle is justified. The inference from the second premise to the conclusion exemplifies IBC:

A belief beats its competitors.
∴ The belief is justified.

An interesting question about IBC arises. Is it stronger than inductions? Can it survive the gauntlet of Hume's argument against inductions? My answer is "No." Hume's argument against induction applies no less to IBC, creating a new philosophical problem that I call "the problem of IBC" according to which IBC is not justifiable. The Humean argument against IBC goes as follows.

IBC is not justified a priori because the negation of the conclusion does not involve a contradiction, i.e., because no contradiction is involved in the statement that the best competitor is not justified. It might be that none of the conceived rival beliefs, including the best competitor, is justified, although the best competitor is better than all the other conceived rival beliefs. This argument echoes Hume's argument that the faculty of reason does not justify the conclusions of inductions because their negations do not involve contradictions.

IBC is not justified a posteriori either because there is no necessary connection between the event that the best competitor beats its competitors and the state of its being justified, i.e., because the event might not be accompanied by the state. After all, we can imagine that the best competitor defeats all objections to it, but that it is not justified. This argument parallels Hume's argument that experience does not justify the conclusions of inductions because in general, cause and effect are not necessarily connected with each other, which implies that two properties that the conclusions of inductions claim are connected with each other are not necessarily connected with each other.

IBC requires a prior belief that I call "the nonexistence principle," according to which it is more likely than not that there are no unconceived objections that would defeat the best competitor. Without the nonexistence principle, it is an illegitimate inference that since the best competitor defeats all the conceived objections to it, it is justified. This claim is similar to Hume's claim that the uniformity principle is required for inductions to be justified, and that without the uniformity principle, the inferences from premises to conclusions of inductions are illegitimate.

The nonexistence principle is not justified a priori because no contradiction is involved in its negation that it is more likely than not that there are unconceived objections that would defeat the best competitor. This argument mirrors Hume's argument that the faculty of reason does not justify the uniformity principle because no contradiction is involved in its negation that nature is not uniform.

The nonexistence principle is not justified a posteriori either because it is circular to use IBC to justify the nonexistence principle. The use of IBC is required to show that the nonexistence principle is justified, i.e., to show that it defeats all objections to it, or to show that it beats its competitors. The nonexistence principle competes with the existence principle that it is more likely than not that there are unconceived objections that would defeat the best competitor. It is not clear how coherentists can show that the nonexistence principle beats the existence principle without relying on IBC. This argument parallels Hume's argument that the uniformity principle is not justified a posteriori because it is circular to use an induction to justify the uniformity principle.

Therefore, IBC is not justifiable, and users of IBC have the burden of justifying the nonexistence principle without relying on IBC. This conclusion is similar to Hume's conclusion that inductions are not justifiable, and that inductivists have the burden of justifying the uniformity principle without relying on an induction.

It would be circular for Lee to use an induction to justify the nonexistence principle. After all, IBC is originally provided to justify the uniformity principle, which is in turn provided to justify inductions. Hence, to use an induction to justify the nonexistence principle amounts to using an induction to justify inductions.

As you can see, Hume's argument against inductions applies no less to IBC. Thus, his argument creates not only the problem of induction but also the problem of IBC. From his point of view, Lee's attempt to justify inductions with IBC is merely an attempt to justify unjustified inferences with another unjustified inferences. It is for this reason that I earlier claimed in Sect. 2.5 above that Lee has to first refute Hume's argument against inductions before constructing the coherentist argument for inductions.

My contention that IBC requires the nonexistence principle is similar not only to Hume's contention that inductions require the uniformity principle, but also to Bas van Fraassen's (1989, pp. 142–143) contention that inference to the best explanation (IBE) requires a principle that I call "the privilege principle," according to which a true hypothesis is more likely to be conceived than unconceived. On van Fraassen's account, the conclusions of IBE are not justified unless the privilege principle is first justified.

Lee endorses van Fraassen's contention that IBE requires the privilege principle according to which "a true hypothesis is included among those we have considered" (Lee, 2022, p. 50). Consequently, Lee would endorse my contention that IBC requires the nonexistence principle. It would be arbitrary for him to say that IBE requires the privilege principle, but that IBC does not require the nonexistence principle.

Coherentists might come up with brilliant ideas to defend IBC vis-à-vis the Humean argument against IBC. Inductivists would welcome such ideas because they could utilize them to defend inductions vis-à-vis Hume's argument against inductions. As a result, they could embrace inductions without relying on the coherentist justification of induction.

2.9 The Default Belief

In this section, I criticize Lee's suggestion that we can justifiably take the uniformity principle as being justified by default.

Appealing to Sellars's (1979, p. 176) insight, Lee states that the uniformity principle "has a default positive justificatory status in our social practice of justification. Accordingly, we are entitled to accept this principle until and unless we are given some positive reasons to think otherwise" (Lee, 2022, p. 41). To put it simply, the uniformity principle is justified by default, and hence we can justifiably believe it until and unless critics provide some reason to disbelieve it.

What are we to make of Lee's Sellarsian claim about the uniformity principle? Imagine that prosecutors claim that it is justified, by default, that defendants are guilty, and thus that they can justifiably believe that defendants are guilty until and unless lawyers provide some reason to disbelieve that defendants are guilty. It does not sound reasonable for prosecutors to say so. Therefore, Lee has the burden of pointing out a relevant difference between his Sellarsian claim about the uniformity principle and prosecutors' Sellarsian claim about defendants.

Moreover, counter-inductivists might suggest that the disuniformity principle is justified by default, and thus that we can justifiably believe it until and unless inductivists provide them with some reasons to disbelieve it. Suppose, for example, that Diana made mistakes whatever they might be, and that counter-inductivists predict that she will no longer make similar mistakes. Counter-inductivists claim that their prediction about her is justified by default, and thus that they can justifiably believe it until and unless inductivists provide them with some reasons to disbelieve it.

Sellarsians can advance the following two claims. First, *the uniformity principle* is justified by default, and we can justifiably believe it until and unless critics provide us with some reason to disbelieve it. Second, *inductions* are justified by default, and we can justifiably believe them until and unless critics provide us with some reason to disbelieve them. Which Sellarsian claim is more reasonable?

The second Sellarsian claim is more reasonable than the first one. The first one is vulnerable to, while the second one is not, to my key objection to Hume's argument against inductions in Sect. 1.3.2 of Chap. 1. My key objection was that Hume confounds a means and an end when he contends that the uniformity principle is required for inductions to be justified. His contention is similar to the contention that the use of a ladder requires being on the roof. Just as it is to confuse a means and an end to say that to be on the roof is required to use a ladder, so it is to confuse a means and an end to say that the uniformity principle is required for inductions to be justified. The unavoidable consequence of confusing the means and the ends is that we can neither have the means nor the ends, i.e., we can neither use the ladder nor be on the roof, and we can neither believe inductions nor the uniformity principle. Lee can avoid the charge of confusing a means and an end by taking inductions as being justified by default rather than taking the uniformity principle as being justified by default.

Lee would retort that he takes the uniformity principle as being justified by default for good reason, viz., "we have relied on UP in order to predict the future on the basis of the past" (Lee, 2022, p. 40). For example, after observing some blue sapphires, we predict that next sapphires will be blue. When we make this prediction, we are relying on the uniformity principle. For this reason, Lee takes the uniformity principle as being justified by default. So would he argue.

However, inductivists would object that the uniformity principle is not required to make the prediction that next sapphires will be blue, and that only the observation of some sapphires is required to make the prediction. The uniformity principle, which implies that the behavior of black holes is uniform, and that overeating and obesity are regularly conjoined, is not required to make the prediction that next sapphires will be blue. We do not need to know about the behavioral pattern of black holes and about the correlation between overeating and obesity to predict that next sapphires will be blue.

Consider also that the uniformity principle is an abstract idea. Only philosophers, who are exposed to Hume's argument against inductions, are familiar with it. It might be that they use it to make predictions. However, it is doubtful that laypeople are also familiar with it, and that they also use it to make predictions. In my view, the uniformity principle is not a required theoretical resource but at best an optional theoretical resource for making predictions.

Furthermore, it is one thing that we use the uniformity principle to make predictions; it is another that it is justified by default. Imagine that fortunetellers claim that they have been using the method of revelation to predict their customers' futures, and that the method of revelation is justified by default. We would say that just because they have been using the method of revelation, it does not follow that it is justified by default. The same holds for the uniformity principle.

2.10 Justification and Probability

In this section, I reply to a possible move that Lee might make in light of my objections to his coherentist justification of induction in the previous sections.

Lee might argue that my objections in the previous sections are built upon a mistaken view of justification that to say that p is justified implies that it is likely to be true. On his account, to say that p is justified does not imply that it is likely to be true, but rather that we ought to accept it to achieve our epistemic goal:

> …epistemic justification (henceforth simply 'justification') is a *normative* concept. We ought to accept justified beliefs, whereas we ought not to accept unjustified beliefs. In addition, justification is a *goal-dependent* concept. To say that a belief is justified is tantamount to saying that it has a favorable status vis-à-vis our epistemic goal. (Lee, 2022, p. 38)

Lee might suggest that the uniformity principle is justified not in the sense that it is likely to be true but rather in the sense that we ought to accept it to achieve our epistemic goal of predicting the future.

How are we to evaluate this possible move? It raises an interesting meta-normative issue of what it means to say that we ought to accept the uniformity principle. Rival metaethicists, such as moral realists, emotivists, and error theorists, would hold rival meta-normative views about the normative claim. I discuss rival meta-normative views in detail from Chaps. 10 to 16 where I defend new metaethical theories and criticize existing metaethical theories, such moral realism and error theory.

In this chapter, I set aside meta-normative issues and raise the following objections to Lee's possible move above. First, why not say that inductions are justified in the sense that we ought to accept them to make predictions? As noted in Sect. 2.10 above, the uniformity principle is at best an optional theoretical resource for making predictions. By contrast, inductions are a required theoretical resource for making predictions. Therefore, it sounds more plausible to say that we ought to accept inductions to make predictions than to say that we ought to accept the uniformity principle to make predictions.

Second, we do not need to dispute Lee's possible suggestion that the uniformity is justified in the sense that we ought to accept it to make predictions. We only need to be careful when we interpret his statement that the uniformity principle is justified. It does not mean that the uniformity principle is likely to be true, but rather that we ought to accept it to make predictions. It is one thing that we ought to accept it to make predictions; it is quite another that it is likely to be true. We should not slide from the former to the latter.

Relatedly, Hume would say that a solution to the problem of induction does not consist in showing that we ought to accept the uniformity principle to make predictions, but rather in showing that inductions are cogent. Inductions are shown to be cogent, insofar as the uniformity principle is shown to be likely to be true. Lee can define 'justified' in any way he wants, and then say that the uniformity principle is justified. It does not follow, however, that he has shown that the uniformity principle is likely to be true.

Lee claims that his "coherentist justification of induction provides a new viable solution to the problems of induction" (Lee, 2022, p. 51). Such a claim creates the burden for him to provide an epistemic reason for preferring the uniformity principle to the disuniformity principle, i.e., the reason for thinking that the uniformity principle is likely to be true. It is not to provide epistemic justification for the uniformity principle for Lee to say that we ought to accept it to make predictions, that "we have no positive reason to think that this principle is false" (Lee, 2022, p. 41), and that the uniformity principle "is indispensable in predicting the future in rational way" (Lee, 2022, p. 41). None of such claims implies that the uniformity principle is likely to be true. We might grant that to say that the uniformity principle beats its competitors implies that it is likely to be true. However, Lee does not even attempt to show that the uniformity principle beats the disuniformity principle, let alone other principles between the uniformity principle and the disuniformity principle.

How would counter-inductivists respond to Lee's foregoing contentions? They would say that we ought to accept the disuniformity principle to make predictions, that we have no positive reason to disbelieve the disuniformity principle, that the disuniformity principle is indispensable in predicting the future in a rational way, and that the disuniformity principle beats its competitors. If Lee thinks that none of such claims implies that the disuniformity principle likely to be true, he should also think that none of his claims above implies that the uniformity principle is likely to be true.

I anticipate that coherentists will now provide their own definition of probability and then claim that on their definition of probability, the uniformity principle is likely to be true. I invite them to imagine how Hume and counter-inductivists would respond to this possible move, and to imagine how defendants would respond to prosecutors, who say to judges, "On the coherentist definition of probability, it is likely that defendants committed crimes. Therefore, judges should convict defendants." To put all this idea differently, coherentists need to show that the degree of our evidence for the uniformity principle is higher than the degree of our evidence for the disuniformity principle rather than to fiddle with the definition of probability. Fiddling with the definition of probability might not impress Hume and counter-inductivists any more than fiddling with the definition of justification would impress them.

2.11 Lee's Use of IBE

In this section, I argue that Lee uses IBE to justify coherentism, and that it is circular for him to do so.

Lee argues that we are not only practically justified but also epistemically justified in believing that all sapphires are blue because "we can answer all objections raised against it in in our social practice of justification" (Lee, 2022, p. 44). In other words, his argument for coherentism is that the "coherentist account explains why the rationality involved in accepting the conclusion of (2) is epistemic rationality rather than practical rationality" (Lee, 2022, p. 44). Note that Lee uses the word "explains" to characterize his attempt to justify coherentism, which indicates that he uses IBE to justify coherentism.

It is circular for Lee to use IBE to justify coherentism. He provides coherentism to justify the uniformity principle, which is in turn provided to justify inductions. To appeal to IBE to justify coherentism amounts to appealing to IBE to justify inductions, which is circular. As specified in Sect. 1.3.1 of Chap. 1, Hume's problem of induction applies not only to enumerative inductions but also to IBE.

Moreover, Lee would not say that the problem of induction applies to enumerative inductions, but not to IBE. After all, he (2022, p. 49) rejects the attempt to solve the problem of induction with the use of IBE, appealing to Ruth Weintraub's (2013) contention that the problem of induction also applies to IBE. On Weintraub's account, IBE requires a principle that might be called "the explainability principle," according to which the world is explainable. However, the explainability principle is not justifiable a priori because no contradiction is involved in the statement that the world is not explainable. Nor is it justifiable a posteriori because it is circular to appeal to experience to justify the explainability principle. It is to use an enumerative induction to say that since past experience has supported the explainability principle, future experience will also support the explainability principle. IBE is originally offered to support enumerative inductions. Therefore, it is circular to use an enumerative induction to justify IBE.

It is prevalent phenomena in philosophy that philosophers openly reject IBE but inadvertently use IBE to establish their positive philosophical hypotheses (Park, 2022, pp. 8–11). For example, van Fraassen (1980, pp. 39–40, Chap. 5) uses IBE to offer positive accounts of phenomena in science, while he (1989, p. 143) constructs arguments against IBE. Larry Laudan (1984, p. 101) and Arthur Fine (1991, p. 83) use IBE to establish their positive philosophical hypotheses, while Laudan (1981, p. 45) and Fine (1991, p. 82) openly claim that it is to beg the question against critics of IBE for scientific realists to use IBE to justify scientific realism. De Ray (2022) openly rejects IBE on the grounds that it requires the explainability principle, but he uses IBE to establish theism (Park, 2023, pp. 482–483). Lee also uses IBE while openly stating that "I do not think that IBE is a legitimate form of reasoning" (Lee, 2022, p. 49).

In my view, it is natural that such phenomena occur in philosophy. IBE is the fundamental rule of inference to establish hypotheses, whether scientific or philosophical. There is no alternative to IBE, which means that if you reject it, there is no way for you to establish a positive hypothesis, whatever that hypothesis might be. This point applies not only to van Fraassen, Laudan, Fine, and Lee but also to Hume. Hume claims that we predict the future because we have the habit of predicting the future. This explanation is not available to Hume, for it falls prey to his argument against inductions (Sect. 1.3.6 of Chap. 1). In short, "Believe little. Explain little" (Park, 2015, p. 28).

2.12 Laws of Nature

In this section, I argue that it is to confuse a means and an end for Lee to say that "If the world is predictable and explainable, some laws of nature or some lawful relations must hold in the world" (Lee, 2022, p. 42).

In my view, the laws of nature support the uniformity principle rather than the other way around, given that the uniformity principle is more general than laws of nature. We first acquire the laws of nature and then infer the uniformity principle. Compare the following two arguments:

(A)
$F = Gm_1m_2/r^2$
$V = IR$
…
∴ Nature is uniform.

(B)
Nature is uniform.
∴ $F = Gm_1m_2/r^2$
∴ $V = IR$
…

In my view, (A) goes hand in hand, while (B) does not, with our cognitive lives. We first observe the world, form laws of nature, and then form the uniformity principle. Accordingly, it is accurate to say that laws of nature support the uniformity principle, but inaccurate to say that the uniformity principle supports laws of nature.

To say that the uniformity principle is required for laws of nature to be justified implies that we should first justifiably believe the uniformity principle before we can justifiably believe laws of nature. To say so is similar to saying that to use a ladder requires that we should first be on the roof. This requirement to use a ladder is so high that no one can meet it, and as a result, no one can use a ladder and be on the roof. Analogously, Lee's requirement to believe laws of nature is so high that no one can meet it, and as a result, no one can believe laws of nature and the uniformity principle. Skepticism about laws of nature and the uniformity principle is an inevitable consequence of Lee's view that the uniformity principle is required for laws of nature to be justified.

2.13 Conclusion

Lee has made an admirable attempt within the coherentist tradition to show that the uniformity principle is justified. His attempt illustrates how coherentists can argue that the uniformity principle is justified. Unfortunately, his achievement does not constitute a solution to the problem of induction.

A solution to the problem of induction consists in showing that the uniformity principle is likely to be true. It is circular for Lee to rely on inductions to justify the uniformity principle and coherentism. Lee faces three difficult tasks: showing that coherentism can withstand Hume's argument against inductions, showing that no induction is involved in the concept of defeat, and justifying coherentism without using an induction.

In addition to coherentism, there are other theories of epistemic justification, such as foundationalism and reliabilism. Is foundationalism or reliabilism useful in solving the problem of induction? A crucial issue is whether the use of an induction is involved in foundationalism and reliabilism. I leave the task of exploring this intriguing issue for a future occasion. For now, I only sum up this chapter with a slogan: "Coherentism cannot solve the problem of induction."

References

Brandom, R. (1994). *Making it explicit*. Harvard University Press.

de Ray, C. (2022). A new epistemological case for theism. *Religious Studies, 58*(2), 379–400.

Fine, A. (1991). Piecemeal realism. *Philosophical Studies, 61*(1–2), 79–96.

Goldman, A. (1979). What is justified belief? In G. S. Pappas (Ed.), *Justification and knowledge: New studies in epistemology* (pp. 1–23). D. Reidel.

Goldman, A., & Beddor, B. (2021). Reliabilist epistemology. In E. N. Zalta (Ed.), *The Stanford encyclopedia of philosophy*. Retrieved from https://plato.stanford.edu/archives/sum2021/entries/reliabilism/

Hume, D. (1993). *An enquiry concerning human understanding*. Hackett Publishing.

Laudan, L. (1981). A confutation of convergent realism. *Philosophy of Science, 48*(1), 19–49.

Laudan, L. (1984). Explaining the success of science: Beyond epistemic realism and relativism. In J. Cushing, C. Delaney, & G. Gutting (Eds.), *Science and reality* (pp. 83–105). University of Notre Dame Press.

Lee, B. (2022). A coherentist justification of induction. *Erkenntnis, 87*(1), 35–52.

Lehrer, K. (2000). *Theory of knowledge*. Westview Press.

Mizrahi, M. (2013). The argument from underconsideration and relative realism. *International Studies in the Philosophy of Science, 27*(4), 393–407.

Park, S. (2015). Explanatory failures of relative realism. *Epistemologia, 38*(1), 16–28.

Park, S. (2017). The uniformity principle vs. the disuniformity principle. *Acta Analytica, 32*(2), 213–222.

Park, S. (2018). The anti-induction for scientific realism. *Grazer Philosophische Studien, 95*(3), 329–342.

Park, S. (2022). *Embracing scientific realism*. Springer.

Park, S. (2023). Inference to the best explanation, naturalism, and theism. *Implicit Religion, 24*, 472–485.

Sellars, W. (1963). *Science, perception and reality*. Ridgeview.

Sellars, W. (1979). *Essays in philosophy and its history*. D. Reidel Publishing.

van Fraassen, B. (1980). *The scientific image*. Oxford University Press.

van Fraassen, B. C. (1989). *Laws and symmetry*. Oxford University Press.

Weintraub, R. (2013). Induction and inference to the best explanation. *Philosophical Studies, 166*(1), 203–216.

Wray, B. (2008). The argument from underconsideration as grounds for anti-realism: A defence. *International Studies in the Philosophy of Science, 22*(3), 317–326.

Chapter 3
Can the Local Theory of Induction Solve the Problem of Induction?

Abstract The local theory of induction asserts that a set of inductions is licensed by a local uniformity principle in a domain, and that the local uniformity principle is justified by other sets of inductions and other local uniformity principles. I object that just as the global theory of induction confounds a means and an end, so the local theory of induction confounds a means and an end. Therefore, just as the global theory of induction leads to skepticism about inductions and the global uniformity principle, so the local theory of induction leads to skepticism about inductions and local uniformity principles.

Keywords Global theory · Induction · Local theory · Material fact · Uniformity principle

3.1 Introduction

There are two different views about induction in the literature. One is the global theory, and the other is the local theory (Norton, 2003). The global theory holds that all (cogent) inductions are justified by the single global principle that might be called "the global uniformity principle" according to which nature is uniform. By contrast, the local theory holds that a set of inductions is justified by a local uniformity principle in a domain, and that the local uniformity principle is justified by the sets of inductions and the local uniformity principles in other domains.

The local theory seems to have an advantage over the global theory. David Hume (1993, pp. 15–25) criticizes the global theory, saying that the global uniformity principle is neither justified a priori nor a posteriori, and hence that no induction is justifiable (Sect. 1.2 of Chap. 1). Thus, the global theory leads to skepticism about inductions. John Norton claims that the local theory, which he refers to as "the material theory" (Norton, 2003, p. 667), gets around the problem of induction. He says, "When we transport the argumentation used to set up the problem of induction to the material theory, it no longer forces the same sort of difficulty" (Norton, 2003, p. 667). In other words, the local theory does not lead to skepticism about inductions.

S. Park, *Induction, Science, and Morality*, Synthese Library 520,
https://doi.org/10.1007/978-3-032-00032-3_3

This chapter aims to show that the alleged advantage of the local theory over the global theory is spurious, and that just as the global theory leads to inductive skepticism, so the local theory leads to inductive skepticism. In Sect. 3.2, I analyze the local theory in detail. In Sect. 3.3, I raise several objections to it. In Sect. 3.4, I reply to a possible move that localists might make to overcome my objections to the local theory. In the end, it will become clear that just as the global theory creates the problem of induction, so the local theory creates the problem of induction.

The present chapter is intended to be useful to those who are interested in whether the local theory shows that inductions are justified, and whether the local theory is preferable to the global theory.

3.2 The Local Theory

In this section, I answer the following questions about the local theory. What does it say? How is it supposed to solve Hume's problem of induction?

The local theory holds that "the admissibility of an induction is ultimately traced back to a matter of fact, not to a universal schema" (Norton, 2003, p. 650). In other words, it is not an argument form but a material fact[1] that makes an inductive argument cogent. Consider, for example, the following two inductions:

(B)
Some samples of the element bismuth melt at 271 °C.
Therefore, all samples of the element bismuth melt at 271 °C. (Norton, 2003, p. 649)

(W)
Some samples of wax melt at 91 °C.
Therefore, all samples of wax melt at 91 °C. (Norton, 2003, p. 649)

These two arguments have the same argument form: some Fs are Gs, and therefore all Fs are Gs. However, (B) is cogent while (W) is not. Why is that so? Norton's answer is that (B) is licensed by the material fact that "all samples of bismuth are uniform" (Norton, 2003, p. 650), while (W) is licensed by no material fact. Why is there no material fact for (W)? There are different kinds of wax. They melt at different temperatures, depending on the mixture of hydrocarbons. Thus, there is no material fact for (W).

On the local theory, a set of inductions in a domain is licensed by its unique material fact. For example, the induction that since Neptune revolved around the sun, it will revolve around the sun is licensed by the material fact that all planets revolve around the sun in the solar system. Another induction that since observed pine trees are green, unobserved pine trees are also green is licensed by another material fact that all pine trees are green. To put it schematically, I_1 is licensed by M_1, I_2 is licensed by M_2, and so forth, where I_1 and M_1 are a set of inductions and a

[1] Norton (2003, *passim*) uses "a material fact" instead of "a local uniformity principle."

material fact, respectively, in a domain, while I_2 and M_2 are another set of inductions and another material fact, respectively, in another domain.

How is the local theory supposed to get around the problem of induction? In other words, how can it be used to show that inductions are justified? Norton answers that "we induce a fact with an induction that is grounded by the facts of a material postulate; these latter facts are justified by inductions that are in turn grounded in the facts of other material postulates" (Norton, 2003, pp. 667–668). To state it otherwise, a set of inductions is licensed by a material fact in a domain, and the material fact is justified by other sets of inductions and other material facts in other domains. Therefore, the set of inductions is justified. To put it schematically, I_1 is licensed by M_1, M_1 is justified by I_2, I_2 is licensed by M_2, M_2 is justified by I_3, and so forth. Therefore, I_1 is justified. Similar localist arguments can be constructed for I_2, I_3, etc. It is an open question how the chain of sets of inductions and material facts ends. It "could end in some sort of circularity," or it could "terminate in brute facts of experience that do not need further justification" (Norton, 2003, p. 668).

Before we evaluate the localist arguments for inductions, we need to be clear about what Norton means by "a material fact." As the quoted sentence above indicates, Norton claims that we can induce a material fact, and that a material fact is justified by the sets of inductions and material facts in other domains. In my view, this use of the word "fact" is unusual. In general, philosophers take a fact to be a piece of the world, but not to be a piece of the mind. They do not take it to be a kind of thing that we can induce and can be justified by a set of inductions or by other facts; they rather take it to be a piece of the world that exists independently of us. It is a category mistake to say that we can induce a piece of the world, and that the piece of the world can be justified by a set of inductions and by other pieces of the world. We can rather induce a general belief in a domain, and the general belief can be justified by the sets of inductions and general beliefs in other domains.

"A material fact" is a misnomer. It gives the impression that we can induce a piece of the world in a domain and justify it with the sets of inductions and pieces of the world in other domains. "A local uniformity principle" better reflects what Norton has in mind than does "a material fact." A local uniformity principle is a general belief about a set of objects or events in a particular domain. We can induce it and justify it with the sets of inductions and local uniformity principles in other domains (at least according to inductivists). Consider Norton's example of a material fact that "all samples of bismuth are uniform" (Norton, 2003, p. 650). This local uniformity principle is a general belief about all bismuth. We can induce it from the observation on bismuth and justify it with the sets of inductions and local uniformity principles in other domains. Thus, Norton's sentence "I_1 is licensed by M_1, and M_1 is justified by I_2" should be interpreted as meaning that a set of inductions is justified by a local uniformity principle in a domain, and the local uniformity principle is justified by another set of inductions in another domain.

In my view, "the material theory" is likewise a misnomer. It confuses readers into thinking that materials in the world justify inductions. For example, bismuth in the world justifies (B). "The local theory of induction" better reflects what Norton has in mind than does "the material theory of induction." The local theory holds that a set of inductions in a domain is licensed by a local uniformity principle, i.e., a local uniformity principle is required for a set of inductions in a domain to be justified.

The local theory contrasts with what I call "the global theory of induction." The global theory holds that all (cogent)[2] inductions in all domains are licensed by the single global principle, viz., the global uniformity principle that nature is uniform. To state it otherwise, the global uniformity principle is required for all inductions in all domains to be justified, and without it no induction would be justified. The global theory is not new in philosophy. Hume criticizes it when he develops the problem of induction. He claims that the global uniformity principle is neither justified a priori nor a posteriori. In consequence, inductions are not justifiable. Hume's criticism against the global theory results in the problem of induction.

From now on, I do not use Norton's terms, "a material fact" and "the material theory." I instead use "a local uniformity principle" and "the local theory." The latter better capture what Norton has in mind than do the former, and that the latter are not misleading at least to those who are new to the debate about the problem of induction.

An interesting question arises. Does the local theory provide a key to solve the problem of induction? Norton's answer is "Yes," while my answer is "No." I criticize Norton's answer and defend my answer in Sect. 3.3 below.

3.3 Critical Responses

3.3.1 Hume's Possible Response

Norton claims that (B) and (W) in Sect. 3.2 above are justified and unjustified, respectively. Hume would disagree with Norton on this count, saying that (B) is just as unjustified as is (W). So I argue in this subsection.

Hume would agree with Norton that the premise of (B) is justified. As we already noted in Sect. 1.2 of Chap. 1, Hume is not skeptical about the premises of inductions. He says, "As to past *Experience*, it can be allowed to give *direct* and *certain* information of those precise objects only" (Hume, 1993, p. 21). However, he would reject Norton's contention that the conclusion of (B) is justified. It is not justified a priori because negating it does not give rise to a contradiction. Nor is it justified a posteriori because there is no necessary connection between the property of being bismuth and the property of melting at 271 °C.

On Hume's account, the global uniformity principle is required for (B) to be justified. However, the global uniformity principle is not justified a priori because its negation is not a contradiction. Nor is it justified a posteriori because it is circular to appeal to an induction to justify it. Consequently, (B) is not justifiable. An implication of this conclusion is that there is no epistemic difference between (B) and

[2] I drop "cogent" from now on for the sake of brevity.

(W), contrary to what Norton thinks. As far as Hume is concerned, both (B) and (W) are "entirely arbitrary" (Hume, 1993, p. 18).

How would Norton respond to Hume's foregoing argument against (B)? Norton would say that what is required for (B) to be justified is not the global uniformity principle but the local uniformity principle that "all samples of bismuth are uniform" (Norton, 2003, p. 650). He would add that the local uniformity principle is justified by other sets of inductions and other local uniformity principles in other domains. It is not circular to provide the sets of inductions and local uniformity principles in other domains to justify the local uniformity principle, although it is circular to provide (B) to justify the local uniformity principle. Therefore, the conclusion of (B) is justified, and the inference of (B) is legitimate.

Interesting questions arise. Would Hume accept Norton's contention that (B) is licensed by the local uniformity principle, and the local uniformity principle is justified by the sets of inductions and local uniformity principles in other domains? I defend a negative answer to this question in the next subsections.

3.3.2 Appealing to Less Powerful Evidence

In this subsection, I raise a sui generis objection to the local theory, an objection that applies to the local theory, but not to the global theory.

Suppose that I_1 and L_1 are a set of inductions and a local uniformity principle in a domain, respectively, and that I_2 and L_2 are another set of inductions and another local uniformity principle in another domain, respectively. Norton's localist argument for I_1 can be put as follows. I_1 is licensed by L_1, L_1 is justified by I_2, I_2 is licensed by L_2, L_2 is justified by I_3, and so forth. Therefore, I_1 is justified. Similar localist arguments can be constructed for I_2, I_3, etc. Therefore, I_1, I_2, and so on are justified.

How are we to evaluate these localist arguments for inductions? Localists provide I_2, L_2, I_3, L_3, and so forth, but not I_1, to justify L_1 because it is circular to provide I_1, but not to provide I_2, L_2, I_3, L_3, and so forth, to justify L_1. This strategy of getting around the problem of circularity comes at the cost of providing less powerful evidence, instead of the most powerful evidence, to justify a local uniformity principle. To use an analogy, localists turn on many less powerful bulbs instead of the most powerful bulb to brighten up a dark room.

Admittedly, L_1 is supported by I_2, L_2, I_3, L_3, and so forth to some extent. However, L_1 is more strongly supported by I_1 than by I_2, L_2, I_3, L_3, and so forth. For example, the local uniformity principle that all planets revolve around the sun is more strongly supported by the induction that since Neptune revolved around the sun, it will move around the sun than by the induction that since observed pine trees are green, unobserved pine trees are green, or by the local uniformity principle that all pine trees are green.

To summarize, the localist strategy to avoid the problem of circularity comes with the cost of providing less powerful evidence, instead of the most powerful evidence, for a local uniformity principle.

3.3.3 *The Medium Theories*

In this subsection, I argue that the local theory is just as vulnerable to the problem of circularity as is the global theory.

Norton claims that the local theory has an advantage over the global theory. On the global theory, all inductions are licensed by the same global principle that nature is uniform. It is circular to provide the global uniformity principle to justify all inductions, and then to provide an induction to justify the global uniformity principle. In consequence, the problem of circularity arises for the global theory. By contrast, the problem of circularity does not arise for the local theory. On the local theory, I_1 is licensed by L_1, and L_1 is justified by all the other sets of inductions and other local uniformity principles in the chain of sets of inductions and local uniformity principles. In other words, when asked to justify I_1, localists provide L_1. When asked to justify L_1, they do not provide I_1, but all the other sets of inductions and other local uniformity principles. It is not circular to do so. So localists claim.

Consider, however, that there can be many medium theories between the local and global theories, and that they can form the following spectrum of theories of induction:

The local theory: I_1 is licensed by L_1. L_1 is justified by I_2. I_2 is licensed by L_2. L_2 is justified by I_3, and so on.

The medium theory 1: I_1 and I_2 are licensed by M_1, where M_1 is a medium uniformity principle. M_1 is justified by I_3 and I_4. I_3 and I_4 are licensed by M_2. M_2 is justified by I_5 and I_6, and so on.

The medium theory 2: I_1, I_2, and I_3 are licensed by M_1. M_1 is justified by I_4, I_5, and I_6. I_4, I_5, and I_6 are licensed by M_2. M_2 is justified by I_7, I_8, and I_9, and so on.

— (Other Medium Theories)

The global theory: All inductions are licensed by the global uniformity principle.

The advocates of the medium theory 1 provide M_1 to justify I_1 and I_2, and they provide all the other sets of inductions and all the other medium uniformity principles to justify M_1. They provide M_2 to justify I_3 and I_4., and they provide all the other sets of inductions and all other medium uniformity principles, including I_1 and M_1, to justify M_2, and so on. The advocates of all the other medium theories do similar things to justify their sets of inductions and medium uniformity principles.

We can run the following reductio ad absurdum against Norton's claim that the local theory is free of the problem of circularity while the global theory is not. If the local theory is free of the problem of circularity, the medium theory 1 is also free of the problem of circularity. We have no reason for thinking that the local theory is free of the problem of circularity while the medium theory 1 is not. If the medium theory 1 is free of the, the medium theory 2 is also free of the problem of circularity. The same holds for all the rest of the medium theories and the global theory. However, the global theory is not free of the problem of circularity, as Hume points out. Therefore, the local theory is not free of the problem of circularity either.

If you are not convinced that medium theories are subject to the problem of circularity, consider the following medium theory. It holds that all inductions can be divided into two groups, I_1 and I_2. I_1 is licensed by M_1, and I_1 is licensed by M_2. There are only four items—I_1, M_1, I_2, and M_2—in the chain of sets of inductions and medium uniformity principles. When asked to justify M_1, the advocates of this medium theory provide I_2 and M_2, but not I_1. When asked to justify M_2, they provide I_1 and M_1, but not I_2. However, it is circular to provide I_2 and M_2 to justify M_1, and then to provide I_1 and M_1 to justify M_2 because that amounts to providing M_1 to justify M_1 and providing M_2 to justify M_2. Similar objections can be raised to all the other medium theories and the local theory in the spectrum of theories of inductions above.

To sum up this subsection, Norton's localist arguments for inductions are just as circular as the globalist argument for inductions. This conclusion will be reinforced in the next subsection.

3.3.4 The Dilemma

In this subsection, I argue that Norton's localist arguments for inductions is subject to the dilemma that I constructed in Sect. 1.3.5 of Chap. 1 against any argument intended to justify inductions.

The dilemma goes as follows. Any argument intended to show that inductions are justified is either deductive or inductive. On the one hand, if it is deductive, its premises assume that inductions are justified. Consequently, the premises need to be justified by new arguments. If new arguments for them are deductive, the problem of infinite regress is in the offing. On the other hand, if the argument is inductive, it is circular to advance it to justify inductions. Therefore, it is in principle impossible to show that inductions are justified.

This dilemma applies to Norton's localist arguments for inductions. They are either deductive or inductive. On the one hand, if they are deductive, the problem of infinite regress is in the offing. On the other hand, if they are inductive, it is circular to provide them to justify inductions. We do not need to determine whether they are deductive or inductive. Either way, none of them shows that an induction is justified, pace Norton.

This dilemma indicates that we can reject Norton's local arguments for inductions without determining whether they are deductive or inductive. However, it is interesting to investigate whether they are deductive or inductive. Consider the localist argument for I_1: I_1 is licensed by L_1, L_1 is justified by I_2, I_2 is licensed by L_2,

L_2 is justified by I_3, and so forth, and thus I_1 is justified. This argument is inductive, given that the premise is compatible with the negation of the conclusion. A contradiction does not arise from affirming the premise and denying the conclusion. In consequence, it is circular for localists to provide the inductive argument to justify I_1.

3.3.5 Begging the Question Against Inductive Skeptics

In this subsection, I argue that inductions feature in Norton's localist arguments for inductions, and thus that his arguments are circular and beg the question against inductive skeptics.

Norton's localist argument for I_1 goes as follows. I_1 is licensed by L_1, L_1 is justified by I_2, I_2 is licensed by L_2, L_2 is justified by I_3, and so forth. Therefore, I_1 is justified. What is wrong with this argument? It is circular to appeal to I_2, I_3, and so forth to justify I_1, given that they are all inductions. The justifiers and the justifiee are the same kind of inferences. Accordingly, it is circular to provide the justifiers to justify the justifiee. So inductive skeptics would say.

Localists would reply that it is not circular to appeal to I_2, I_3, and so forth to justify I_1, although they are the same kind of inferences, because they are different inductions with different contents. For example, I_1 is an induction about planets, I_2 is an induction about pine trees, and so on. Providing I_2, I_3, and so on to justify I_1 is different from providing I_1 to justify I_1, which is obviously circular. It follows that it is not circular to provide I_2, I_3, and so forth to justify I_1.

This reply sounds reasonable to me. However, it runs counter to the standard Humean objection to inductivists' attempt to justify inductions. As noted in Sect. 1.3.4 of Chap. 1, inductivists advance the following second-order argument to justify inductions:

Inductions have worked in the past.
∴ They will also work in the future.

This second-order induction seems to show that first-order inductions are justified. However, Humean skeptics object that it is circular to provide the second-order induction to justify first-order inductions because they are the same kind of inferences.

In my view, if it is not circular for localists to provide I_2, I_3, and so forth to justify I_1, it is not circular either for second-order inductivists to provide the second-order induction to justify first-order inductions. After all, just as I_1, I_2, I_3, and so on are different inductions with different contents, so the second-order induction and first-order inductions are different inductions with different contents. If it is circular for second-order inductivists to provide the second-order induction to justify first-order inductions, it is also circular for localists to provide I_2, I_3, and so on to justify I_1. In short, the localist argument for I_1 and the second-order inductivist argument for first-order inductions are in the same boat.

An interesting implication of this observation is that second-order inductivists do not need to choose the localist argument for I_1 over the second-order inductivist argument for first-order inductions to overcome the problem of induction. Moreover, it sounds more plausible to say that the second-order induction is required for first-order inductions to be justified than to say that a local uniformity principle is required for a set of first-order inductions to be justified. After all, to say that a general belief is required for particular beliefs to be justified is similar to saying that to be on the roof is required to use a ladder, as noted in Sect. 1.3.2 of Chap. 1.

Now, what would inductive skeptics say about the localist argument for I_1 and the second-order inductivist argument for inductions? They would say that if I_1 needs to be justified because it is an induction, I_2, I_3, I_4, and so on also need to be justified because they are inductions. It achieves nothing to provide unjustified inferences to justify an unjustified inference. The same holds for the second-order induction. If first-order inductions need to be justified because they are inductions, the second-order induction also needs to be justified because it is an induction. It achieves nothing to provide an unjustified inference to justify unjustified inferences. In short, it is to beg the questions against inductive skeptics for localists and second-order inductivists to assume that inductions have supporting power.

Let me draw a philosophical lesson. Inductivists need to refute Hume's argument against inductions before they construct an inductive argument. If they do not refute it, inductive skeptics will come along and object that their inductive argument is not justified. Let me remind inductivists that Hume's argument against inductions is vulnerable to my six objections in Sect. 1.3 of Chap. 1.

3.3.6 Begging the Question Against Inductivists

In this subsection, I argue that Norton's localist arguments for inductions beg the question against inductivists.

Hume and inductivists have opposing views about inductions. Hume affirms, while inductivists deny, that the global uniformity principle is required for inductions to be justified (Sect. 1.3.2 of Chap. 1). Their opposing views can be illustrated by the following two arguments, (I) and (G):

(I)
Neptune revolved around the sun.
∴ Neptune will revolve around the sun.

(G)
Neptune revolved around the sun.
Nature is uniform.
∴ Neptune will revolve around the sun.

Note that the global uniformity principle features in (G), but not in (I). According to Hume, (I) is not justified while (G) is justified. According to inductivists, by contrast, (I) is justified, i.e., the premise is powerful enough to justify the conclusion, and the premise does not need the global uniformity principle to justify the conclusion.

There is a noteworthy difference between (I) and (G): (I) is an induction while (G) is a deduction. Accordingly, Hume begs the question against inductivists when he claims that the global uniformity principle is required for (I) to be justified. After all, this claim implies that an induction is not justified while a deduction is justified. He assumes that an induction is not justified in his attempt to show that inductions are not justifiable (Sect. 1.3.2 of Chap. 1).

Hume would say that the conclusion of (I) is neither justified a priori nor a posteriori. It is not justified a priori because its negation is not a contradiction. It is not justified a posteriori because the property of being Neptune and the property of revolving around the sun are not necessarily connected each other (Sect. 1.2 of Chap. 1).

However, this objection to (I) is built up the dubious assumption that a justified belief is *necessarily* true, and it would be rejected by contemporary epistemologists who believe that a justified belief is *probably* true. Moreover, the objection overlooks the possibility that the faculty of reason and experience might jointly justify the conclusion of (I) (Sect. 1.3.1 of Chap. 1).

How about Norton's localist theory? In my view, it also begs the question against inductivists. Localists and inductivists have opposing views on inductions. Localists affirm, while inductivists deny, that a local uniformity principle is required for an induction to be justified. Their opposing views can be illustrated by the following two arguments, (I) and (L):

(I)
Neptune revolved around the sun.
∴ Neptune will revolve around the sun.

(L)
Neptune revolved around the sun.
Planets revolve around the sun.
∴ Neptune will revolve around the sun.

Note that a local uniformity principle figures in (L), but not in (I). According to localists, (I) is not justified while (L) is justified. According to inductivists, by contrast, (I) is also justified, i.e., the premise is powerful enough to justify the conclusion, and the premise does not need the local uniformity principle to justify the conclusion.

There is a noteworthy difference between (I) and (L). (I) is an induction while (L) is a deduction. Accordingly, localists beg the question against inductivists when localists claim that the local uniformity principle is required for (I) to be justified. Inductivists would reject localists' claim that (I) is not justified without the local uniformity principle, i.e., that the local uniformity principle is required for (I) to be justified.

What justifies localists' assumption that (I) is not justified, but that (L) is justified? What would be the reason for thinking that the premise of (I) alone cannot justify the conclusion? It would be problematic for localists to answer that (I) is not justified because the negation of the conclusion is not a contradiction, and that the property of being Neptune and the property of revolving around the sun are not necessarily connected with each other. After all, no inductivist claims that a justified belief is necessarily true, and that the faculty of reason and experience separately justify the conclusion of (I).

It is my diagnosis that just as globalists beg the question against inductivists when they claim that the global uniformity principle is required for inductions to be justified, so localists beg the question against inductivists when they claim that a local uniformity principle is required for a set of inductions to be justified in a domain. Just as Hume has the burden of justifying his dubious assumption that no induction is justified without the global uniformity principle, so Norton has the burden of justifying his dubious assumption that a set of inductions is not justified without a local uniformity principle.

3.3.7 Confounding a Means and an End

In this subsection, I argue that just as the global theory confuses a means and an end, so the local theory confuses a means and an end.

Globalists suggest that the global uniformity principle is required for inductions to be justified. Consider the relationship between the following induction and the global uniformity principle:

(I) Since Neptune revolved around the sun, it will revolve around the sun.
(U) Nature is uniform.

Which is a means, and which is an end? In my view, the induction is a means, and the global uniformity principle is an end. The global uniformity principle is an end that we can arrive at by means of the induction and other inductions. As a consequence, it is accurate to say that inductions are required for the global uniformity principle to be justified, but inaccurate to say that the global uniformity principle is required for inductions to be justified (Sect. 1.3.2 of Chap. 1).

To say that the global uniformity principle is required for inductions to be justified is to confuse the direction of the requirement between inductions and the global uniformity principle. It is similar to the suggestion that being on the roof is required for using a ladder. In reality, using a ladder is a means, and being on the roof is an end. However, if you confound the means and the end, and say that being on the roof is required for using a ladder, you can neither use a ladder nor be on the roof. Similarly, if you confuse the direction of the requirement between inductions and the global uniformity principle, and say that the global uniformity principle is required for inductions to be justified, you can justifiably believe neither inductions nor the global uniformity principle. Skepticism about both inductions and the global uniformity principle is an unavoidable consequence of confusing the means and the end (Sect. 1.3.2 of Chap. 1).

Just as the global theory confuses a means and an end, so the local theory confuses a means and an end. Consider the following induction (I) and local uniformity principle (L):

(I) Since Neptune revolved around the sun, it will revolve around the sun.
(L) Planets revolve around the sun.

Which is a means, and which is an end? In my view, the induction is a means, and the local uniformity principle is an end. The local uniformity principle is an end that we can arrive at by means of the induction and other inductions about planets in a solar system. In consequence, it is accurate to say that a set of inductions about planets is required for a local uniformity principle to be justified, but inaccurate to say that the local uniformity principle is required for a set of inductions about planets to be justified.

To say that a local uniformity principle is required for a set of inductions to be justified is to confuse the direction of requirement between them. It is similar to the suggestion that being on the roof is required for using a ladder. If you confuse the direction of the requirement between a set of inductions and a local uniformity principle, and say that the local uniformity principle is required for the set of inductions to be justified, you can justifiably believe neither the set of inductions nor the local uniformity principle. Skepticism about both the set of inductions and the local uniformity principle is the inevitable consequence of confounding the means and the end, pace Norton.

The analogy of the roof and a ladder is useful to show that Norton's localist argument for I_l leads to skepticism about a set of inductions and a local uniformity principle. Imagine that there are many ladders, l_1, l_2, l_3, and so on, and that there are many roofs, r_1, r_2, r_3, and so on. Each roof has its own ladder: r_1 has l_1, r_2 has l_2, r_3 has l_3, and so on. There is a condition for using a ladder: to be on r_2 is required for you to use l_1, to be on r_3 is required for you to use l_2, and so on. Can you use l_1 and be on r_1? My answer is "No." You cannot use l_1 to be on r_1 because you should first be on r_2 to use l_1. Nor can you use l_2 to be on r_2 because you should first be on r_3 to use l_2. The same holds for using other ladders. As a result, you can use none of the ladders and can be on none of the roofs.

Just as Hume creates the problem of induction with the global theory, so Norton creates the problem of induction with the local theory. Hume contends that the global uniformity principle is required for inductions to be justified. As a result, he can justifiably believe neither inductions nor the global uniformity principle. The same holds for Norton. He contends that a local uniformity principle is required for an induction to be justified. As a result, he can justifiably believe neither an induction nor a local uniformity principle. Accordingly, it does not matter whether you say that a local uniformity principle is required for a set of inductions to be justified in a domain, or that the global principle is required for all inductions in all domains to be justified. Insofar as you confound the means and the ends, you cannot avoid skepticism about both the means and the ends, i.e., both they are forever beyond your reach.

The confusion of a means and an end is apparent in Norton's language. He writes, "the more we know, the better we can infer inductively" (Norton, 2003, p. 664), "the mere fact of learning more will augment their inductive powers automatically" (Norton, 2003, p. 664), and "In order to learn a fact by induction, the material theory says that we must already know a fact, the material postulate that licenses the induction" (Norton, 2003, p. 666). In my view, all these statements are similar to the statement that we should already be on a roof to use a ladder. How can we already know about unobserved objects without first using an induction from observed objects to them? How can we already be on a roof without first using a ladder? It is incomprehensible to me how we already know that a local uniformity principle is true without first relying on an induction.

3.3.8 Counter-Induction

Counter-inductivists (Salmon, 1957) construct a counter-induction to drive home the problem of induction (Sect 1.4 of Chap. 1). In this subsection, I sketch how local counter-inductivists would respond to Norton's local inductivist arguments for inductions. This possible response from local counter-inductivists will strengthen my case that the local theory creates rather than solves the problem of induction.

How do global inductivists and global counter-inductivists differ from each other? Just as global inductivists rely on the global uniformity principle to predict the future, so global counter-inductivists rely on the global disuniformity principle (Park, 2017, 2018, 2022, pp. 76–79) to predict the future. The global uniformity principle holds that nature is uniform, while the global disuniformity principle holds that nature is disuniform. As a result, global inductivists and global counter-inductivists draw opposite conclusions from the same premise. For example, from the same premise that Neptune revolved around the sun, inductivists draw the conclusion that it will revolve around the sun while counter-inductivists draw the opposite conclusion that it will not revolve around the sun.

Local inductivists and local counter-inductivists would use a local uniformity principle and a local disuniformity principle, respectively, to predict the future in a domain. For example, local inductivists use the local uniformity principle that planets revolve around the sun while local counter-inductivists use the local disuniformity principle that planets will no longer revolve around the sun. As a result, just as global inductivists and global counter-inductivists draw opposite conclusion from the same premise, so local inductivists and local counter-inductivists draw opposite conclusions from the same premise.

How would local inductivists attempt to break the tie between the two opposing predictions about Neptune? They would provide another local uniformity principle that all pine trees are green, another local uniformity principle that apples thrown upwards fall down, and so forth, to reinforce their prediction that Neptune will revolve around the sun. On the local inductivist account, the more such local uniformity principles we have, the stronger reason we have for predicting that Neptune will revolve around the sun.

In response, local counter-inductivists would provide a local disuniformity principle that unobserved pine trees are not green, another local uniformity principle that apples thrown upwards will no longer fall down, and so on, to reinforce their prediction that Neptune will no longer revolve around the sun. On the local counter-inductivist account, the more such local disuniformity principles we have, the stronger reason we have for predicting that Neptune will not revolve around the sun.

Local inductivist arguments for inductions and local counter-inductivist arguments for counter-inductions have a similar structure. The local inductivist argument for I_1 goes as follows. I_1 is licensed by L_1, L_1 is justified by I_2, I_2 is licensed by L_2, L_2 is justified by I_3, and so forth. Therefore, I_1 is justified. The local counter-inductivist argument for C_1 goes as follows. Suppose that C_1 is a set of counter-inductions, that D_1 is a local disuniformity principle, that C_2 is another set of counter-inductions, and that D_2 is another local disuniformity principle. On the counter-inductivist account, C_1 is licensed by D_1, D_1 is justified by C_2, that C_2 is licensed by D_2, D_2 is justified by C_3, and so on. Therefore, C_1 is justified.

This confrontation between the local inductivist argument for I_1 and the local counter-inductivist argument for C_1 motivates me to construct the following dilemma against local inductivists, such as Norton. The local counter-inductivist arguments for counter-inductions either show or do not show that counter-inductions are justified. On the one hand, if they show that counter-inductions are justified, inductions are not justified, given that counter-inductions conflict with inductions. On the other hand, if they do not show that counter-inductions are justified. Neither do Norton's local inductivist arguments for inductions show that inductions are justified, given that the local counter-inductivist arguments for counter-inductions mirror the local inductivist arguments for inductions. Either way, the local inductivist arguments for inductions do not show that inductions are justified.

A philosophical lesson from this conclusion is that it does not matter whether you attempt to justify inductions locally or globally. You cannot justify inductions, i.e., you cannot solve the problem of induction.

3.4 Objection and Reply

In this section, I reply to a possible objection that I have misunderstood Norton's local theory, and that my criticisms against it commit the straw man fallacy.

Local inductivists might object that the local theory does not claim that a local uniformity principle is required for an induction to be justified, and that it rather claims that a material *fact* is required for an induction to be justified. A local uniformity principle is a general belief, and thus it is a piece of the mind, while a material fact is a piece of the world. It might beg the question against inductivists and might confuse a means and an end to say that a local uniformity principle is required for a set of inductions to be justified. However, it does not commit those mistakes to say that a material fact is required for a set of inductions to be justified. After all, it is legitimate to say that if a material fact did not exist in the world, we would not form a set of inductions, i.e., that there is a causal relationship between a material fact in the world and a set of inductions in our minds. For example, if the material fact that planets revolve around the sun did not exist in the world, we would not form the induction in our minds that since Neptune revolved around the sun, it will revolve around the sun. Consequently, the material fact justifies the induction.

My immediate response to this possible move is to say that it does not mesh well with what Norton says about a material fact. He says, for example, that "we induce a fact with an induction that is grounded by the facts of a material postulate; these latter facts are justified by inductions that are in turn grounded in the facts of other material postulates" (Norton, 2003, pp. 667–668). It is incomprehensible to me that we can induce a piece of the world, or that a piece of the world can be justified by a set of inductions or by other pieces of the world. It seems to be a stretch to interpret Norton's term "a material fact" as referring to a piece of the world.

However, we might grant for the sake of argument that for Norton, a material fact is a piece of the world. An immediate objection to his local inductivist arguments for inductions is to ask how we know that a material fact exists. For example, how do we know that planets revolve around the sun? Local inductivists would answer that we rely on sets of inductions about non-planets and material facts of non-planets. However, this answer involves providing less powerful evidence, instead of the most powerful evidence, to justify a belief about the general fact. The most powerful evidence for the belief about the general fact is a set of inductions about planets, but not the other sets of inductions about non-planets and other material facts of non-planets.

Moreover, it is still to confound a means and an end to suggest that the knowledge about a material fact, a piece of the world, is required for a set of inductions to be justified. Consider the material fact that planets revolve around the sun, and the induction that since Neptune revolved around the sun, it will revolve around the sun. We come to identify the material fact that planets revolve around the sun with the use of the induction and other inductions about planets rather than the other way around. To state it otherwise, the inductions are the means to arrive at the material fact rather than the other way around. To suggest that we should first recognize the material fact before we use the inductions about planets is similar to the suggestion that we should first be on the roof to use a ladder.

To sum up this section, shifting the meaning of "a material fact" from a local uniformity principle to a piece of the world does not help the local theory at all. Whether a material fact is a local uniformity principle or a piece of the world, the local theory leads to skepticism about both inductions and local uniformity principles.

3.5 Conclusion

Just as the global theory asserts that the global uniformity principle is required for all inductions in all domains to be justified, so the local theory asserts that a local uniformity principle is required for a set of inductions to be justified in a domain. Just as the global theory is the source of skepticism about inductions, so the local theory is the source of skepticism about inductions. I dare say that once you endorse either the global theory or the local theory, you cannot avoid skepticism about inductions.

Some readers would be tempted to object that since I rejected both the global and local theories, I have the burden of providing an alternative theory of induction. They might ask, "What is your positive strategy to justify inductions?" My response to this challenge is to challenge it with the question that I previously asked: "Why do you think that inductions need to be justified?" (Sect. 1.3.5 of Chap. 1) It is wrong to answer, "Because their conclusions are neither justified a priori nor a posteriori." This answer from Hume is not adequate, as we have seen in Sect. 1.3.1 of Chap. 1 and in Sect. 3.3.6 of this chapter.

Let me finish this chapter with the following slogan: "Going local doesn't help."

References

Hume, D. (1993). *An enquiry concerning human understanding*. Hackett Publishing.

Norton, J. (2003). A material theory of induction. *Philosophy of Science, 70*(4), 647–670.

Park, S. (2017). The uniformity principle vs. the disuniformity Principle. *Acta Analytica, 32*(2), 213–222.

Park, S. (2018). The anti-induction for scientific realism. *Grazer Philosophische Studien, 95*(3), 329–342.

Park, S. (2022). *Embracing scientific realism*. Springer.

Salmon, W. (1957). Should we attempt to justify induction? *Philosophical Studies, 8*(3), 33–48.

Chapter 4
The Excessive Requirements on the Use of IBE

Abstract As a result of using inference to the best explanation (IBE), we gain confidence in the belief that a true hypothesis is more likely to be conceived than unconceived, and in the belief that the world is explainable. Therefore, the use of IBE is required for these two beliefs to be justified rather than the other way around. However, van Fraassen's argument from a bad lot and de Ray's argument from a brute fact presuppose erroneously that the two beliefs are required for IBE to be justified. This false presupposition results in excessive requirements on the use of IBE, in skepticism about our best scientific theories, and in backfires on van Fraassen's use of IBE and de Ray's use of IBE.

Keywords Bad lot · Brute fact · de Ray · Inference to the best explanation · van Fraassen

4.1 Introduction

Inference to the best explanation (IBE) is widely used in science, in philosophy, and in our daily lives. We consider rival hypotheses that explain puzzling phenomena and then choose the one that best explains them. Suppose, for example, that you were denied a promotion from assistant professor to associate professor. You compare the hypothesis that you were not qualified to be promoted with the hypothesis that the committee members were unprofessional. You choose the latter on the grounds that you do not get along with them, and on the grounds that your predecessors, who were less qualified than you, were promoted from assistant professors to associate professors.

Bas van Fraassen (1989, p. 143) constructs the argument from a bad lot against IBE. Ruth Weintraub (2013, p. 213) and Christophe de Ray (2022) construct an argument against IBE that I call "the argument from a brute fact." I unpack these two arguments in Sect. 4.2. In Sect. 4.3, I point out that van Fraassen and de Ray use IBE themselves to justify their positive philosophical views, and thus that their critiques of IBE backfire on their own uses of IBE and lead to skepticism about their

S. Park, *Induction, Science, and Morality*, Synthese Library 520,
https://doi.org/10.1007/978-3-032-00032-3_4

positive philosophical views. In Sect. 4.4, I expose three intrinsic problems with the arguments from a bad lot and a brute fact.

This chapter is intended to be useful to those who aim to gain some insight into how IBE works, how IBE expands our intellectual horizons, and whether the requirements that van Fraassen, Weintraub, and de Ray impose on the use of IBE are reasonable.

4.2 The Arguments from a Bad Lot and a Brute Fact

In this section, I unfold the arguments from a bad lot and a brute fact. Readers will see that the two arguments impose excessive requirements on the use of IBE.

The argument from a bad lot consists of a premise and a conclusion. The premise holds that H, the best of conceived rival hypotheses, "may be the best of a bad lot" (van Fraassen, 1989, p. 143). In other words, it is possible that all of the rival hypotheses that users of IBE have considered might be false. The conclusion holds that "the best of set *X* will be more likely to be true than not, requires a prior belief that the truth is already more likely to be found in *X*, than not" (van Fraassen, 1989, p. 143). To put it differently, before concluding that H is true, users of IBE should show that a true hypothesis is more likely to be found in the collection of the rivals that they considered than in the collection of the rivals that they did not. I call this prior belief "the privilege principle" because it implies that users of IBE are epistemically privileged to discover truths. The argument from a bad lot comes down to the request that users of IBE justify the privilege principle without using IBE. Moti Mizrahi (2020, pp. 54, 128) and K. Brad Wray (2021, p. 246) take this argument to be a strong objection to scientific realism, the view that H is true.

Let me apply the argument from a bad lot to the example of the promotion above. The best hypothesis that the committee members were unprofessional might be the best of a bad lot. Accordingly, before you conclude that the committee members were unprofessional, you have to justify the privilege principle, according to which the true hypothesis explaining why you were denied the promotion is more likely to be conceived than unconceived, i.e., that the set of the rivals that you considered is more likely to include the true hypothesis than not.

How can we go about justifying the privilege principle? In my view, this question can be answered with the use of the following analogy. If we aim to justify the general belief that most Koreans are Christians, we should randomly select some Koreans from the population of all Koreans and then investigate whether they are Christians. If most of them are Christians, we can conclude that the general belief is true. Similarly, if we aim to justify the privilege principle, we should randomly select some hypotheses from the population of all hypotheses in our intellectual enterprises and then investigate whether they were chosen from good lots or bad lots. If most of them were chosen from good lots, we can conclude that the privilege principle is true.

Let me turn to the argument from a brute fact, which also consists of a premise and a conclusion. The premise holds that a fact might be brute, i.e., that a fact might not have an explanation. For example, we "can imagine that nothing at all explains the appearance of pellets on the floor and strange marks on the cheese this morning" (de Ray, 2022, p. 382). The conclusion holds that the use of IBE requires the prior belief that the world is (overall) explainable. For example, to say that there are pellets on the floor because there was a mouse on the floor requires the prior belief that most phenomena have explanations. Weintraub states, "Every IBE argument assumes that nature is explicable" (Weintraub, 2013, p. 213). De Ray states, "relying on IBE does seem to require at the very least the belief that states of affairs at least generally have explanations for their obtaining" (de Ray, 2022, p. 383). Let me call this prior belief "the explainability principle" because it asserts that the world is explainable. Thus, the argument from a brute fact boils down to the request that users of IBE justify the explainability principle without using IBE.

Let me apply the argument from a brute fact to the example of the promotion in Sect. 4.1 above. It might be a brute fact that you were denied the promotion, i.e., the fact that you were denied the promotion might not have an explanation. Accordingly, before you conclude that the committee members were unprofessional, you have to justify the explainability principle that most phenomena, e.g., why the sky is blue and why a neutron has mass, have explanations.

How can we go about justifying the explainability principle? In my view, we should justify the explainability principle in the way we justify the privilege principle. Specifically, we should randomly select some phenomena from the population of all phenomena in the world and then investigate whether they have explanations. If most of them have explanations, we can conclude that the explainability principle is true.

Just as the argument from a bad lot claims that users of IBE cannot conclude that H is true before they justify the privilege principle without using IBE, so the argument from a brute fact claims that users of IBE cannot conclude that H is true before they justify the explainability principle without using IBE. In my view, both the arguments from a bad lot and a brute fact impose excessive requirements on the use of IBE, thereby inviting the situation where no one can legitimately use IBE.

4.3 Backfires

In this section, I assume that the arguments from a bad lot and a brute fact are acceptable and then elucidate their disastrous implications against van Fraassen's and de Ray's own uses of IBE.

Van Fraassen (1980, Chap. 5) uses IBE to establish his theory of scientific explanation called "the contextual theory" (Park, 2019, p. 91). To put it briefly, van Fraassen claims that his theory can account for asymmetries and rejections, phenomena that occur in science. By contrast, his rival theories, such as Carl Hempel's (1966) deductive-nomological model, cannot account for these phenomena.

Therefore, the contextual theory is "the correct account" (van Fraassen, 1980, p. 112), and "An explanation is an answer to a why-question" (van Fraassen, 1980, p. 134).

Van Fraassen has the burden of meeting the excessive requirement that he himself imposes on the use of IBE. Specifically, before he concludes that a scientific explanation is a response to a why-question, he has to justify the privilege principle that a true theory of scientific explanation is more likely to be found in the collection of the rivals that he considered than in the collection of the rivals that he did not. He should meet this request without using IBE.

Van Fraassen would run into the problem of infinite regress, if he used IBE to justify the privilege principle, i.e., if he argued that the privilege principle is true because it best explains certain phenomena, whatever they might be. After all, to argue so would create the burden of justifying the second-order privilege principle that the first-order privilege principle is more likely to be chosen from a good lot than from a bad lot. To use IBE to justify the second-order privilege principle would create the burden of justifying the third-order privilege principle, and so forth. As a result, van Fraassen would never be able to justify the first-order privilege principle.

Alternatively, van Fraassen might randomly select some hypotheses from the population of all hypotheses in our intellectual enterprises and then investigate whether they were chosen from good lots or bad lots. If most of them were chosen from good lots, he can conclude that the privilege principle is true, and hence that the contextual theory is true. He should not use IBE to show that most of the selected phenomena have explanations. If he used IBE, he would run into the problem of infinite regress.

De Ray also uses IBE to establish the theist hypothesis that God causes the explainability principle in our minds (Park, 2023, pp. 482–483). He compares the theist hypothesis with the naturalist hypothesis. According to the naturalist hypothesis, the general fact that most phenomena have explanations causes the explainability principle in our minds. De Ray claims that the general fact cannot cause the explainability principle in our minds, but that God can cause the explainability principle in our minds (de Ray, 2022, p. 393). For this reason, he rejects the naturalist hypothesis and accepts the theist hypothesis. He says, "The proposed aetiology is therefore plausible *given theism*. In contrast, I have argued that there are no aetiologies of the intelligibility intuition which are plausible *given naturalism*" (de Ray, 2022, p. 393). In plain English, the theist hypothesis better explains why we hold the explainability principle than the naturalist hypothesis does. Note that de Ray uses IBE to justify the theist hypothesis.

De Ray has the burden of meeting the excessive requirement that he himself imposes on the use of IBE. Specifically, before he concludes that God produces the explainability principle in our minds, he should justify the explainability principle. He should meet this request without using IBE. He would run into the problem of infinite regress, if he used IBE to justify the explainability principle, i.e., if he argued that the explainability principle is true on the grounds that it best explains certain phenomena, whatever they might be. After all, to argue so would require the second-order explainability principle. If he were to use IBE again to justify the second-order explainability principle, he would again have the burden of justifying the third-order explainability principle, and so on. As a result, he would never be able to justify the first-order explainability principle.

Alternatively, de Ray might randomly select some phenomena from the population of all phenomena in the world, and then show that most of the selected phenomena have explanations. If most of them have explanations, he can conclude that the privilege principle is true, and hence that his theist hypothesis is true. He should not use IBE to show that most of the selected phenomena have explanations. If he used IBE, he would run into the problem of infinite regress.

We can turn de Ray's requirement against van Fraassen's use of IBE. Van Fraassen's use of IBE requires the explainability principle, which implies that before he concludes that a scientific explanation is a response to a why-question, he should justify the explainability principle without using IBE. If he used IBE to justify the explainability principle, he would run into the problem of infinite regress. If his use of the random sampling method involved the use of IBE, the problem of infinite regress would also arise.

We can also turn van Fraassen's requirement against de Ray's use of IBE. De Ray's use of IBE requires the privilege principle, according to which the true hypothesis explaining why we hold the explainability principle is more likely to be in the collection of the rivals that he considered than in the collection of the rivals that he did not. He has to justify the privilege principle before he concludes that God causes the explainability principle in our minds. He should not use IBE to justify the privilege principle. If he used IBE to justify the privilege principle, he would run into the problem of infinite regress. If his use of the random sampling method involved the use of IBE, the problem of infinite regress would also arise.

4.4 Intrinsic Problems with the Two Arguments

4.4.1 Drawing the Line

In Sect. 4.2 above, I argued that van Fraassen and de Ray might use the random sampling method to justify the privilege principle and the explainability principle, respectively, and that if their uses of random sampling methods involved the uses of IBE, the problem of infinite regress would arise. In this subsection, I raise another difficulty against using the random sampling method.

Suppose again that you were denied the promotion, hold the hypothesis that the committee members were unprofessional, and aim to justify the explainability principle with the random sampling method. On close examination, you should not select some phenomena from the population of all phenomena in the world, but rather from the population of the phenomena that are similar to the phenomenon that you were denied the promotion. Suppose, for example, that you have a colleague who is also an assistant professor, does not get along with the committee members, is better qualified than your predecessors, but is denied the promotion. This fact is similar to the fact that you were denied the promotion, and if the former has an explanation, it is likely that the latter also has an explanation. Consequently, facts like the former should be included in the population of the phenomena from which you select the samples.

By contrast, suppose that God exists. This fact is dissimilar to the fact that you were denied the promotion. Just because the former does not have an explanation, it does not follow that the latter does not have an explanation either. Accordingly, facts like the former should be excluded from the population of the phenomena from which you select the samples. If facts like the former are included in the population, and if some of them are selected, they would only bias your sample and contaminate your judgment about whether the fact that you were denied the promotion has an explanation and whether the committee members were unprofessional.

Between these two extreme phenomena, there is a spectrum of phenomena, such as the phenomena that 4 + 5 = 9, that the sky is blue, and that a job candidate was not offered a position. The existence of the spectrum triggers the following question. How should we delineate the population from which we can select some phenomena to conclude that the committee members were unprofessional? Where should we draw the line between the phenomena that should be included in the population and the phenomena that should be excluded from the population? I leave the task of answering this question to those who are sympathetic to the argument from a brute fact.

At first sight, the argument from a bad lot does not seem to suffer from a similar problem. As noted in Sect. 4.2 above, the argument from a bad lot implies that before you conclude that the committee members were unprofessional, you should justify the privilege principle, according to which the true hypothesis explaining why you were denied the promotion is more likely to be conceived than unconceived. The privilege principle concerns the true hypothesis, but not other hypotheses. Accordingly, the scope of the argument from a bad lot seems to be precise.

On close examination, however, the argument from a bad lot does suffer from a similar problem. Suppose that you have a colleague who is also an assistant professor, does not get along with the committee members, is better qualified than your predecessors, but is denied a promotion to associate professor. Under such circumstances, the true hypothesis explaining why your colleague was denied the promotion is similar to the true hypothesis explaining why you were denied the promotion, so the former can replace the latter in the argument from a bad lot leveled at your use of IBE. With the replacement, the argument from a bad lot reads that your use of IBE requires the privilege principle, according to which the true hypothesis explaining why your colleague is denied the promotion is more likely to be chosen from a good lot than from a bad lot. If your colleague somehow established the privilege principle, you could appeal to his accomplishment and believe that the committee members were unprofessional.

What if you have another colleague who is an associate professor and was denied a promotion to full professor? Can the true hypothesis explaining why he failed to be promoted replace the true hypothesis explaining why you failed to be promoted?

What about the true hypothesis explaining why the sky is blue? What about the true hypothesis explaining why 4 + 5 = 9? What about the true hypothesis explaining why God exists? Obviously, the true hypothesis explaining why God exists cannot replace the true hypothesis explaining why you were denied the promotion. Even if theologians investigating why God exists somehow established the privilege principle according to which the true hypothesis explaining why God exists is more likely to be conceived than unconceived, you could not appeal to their accomplishment and believe that the committee members were unprofessional.

As a consequence, we have an unanswered question: where should we draw the line between the true hypotheses that can replace the true hypothesis explaining why you were denied the promotion and the true hypotheses that cannot? This question should be answered by those who are sympathetic to the argument from a bad lot. I admit, though, that the question is not a fatal challenge to the argument from a bad lot, and that it is merely a request that the argument from a bad lot should be sharpened.

In the next subsections, however, I raise more substantial objections to the arguments from a bad lot and a brute fact.

4.4.2 Possibility vs. Probabilistic Statements

The premise of the argument from a bad lot is, as noted in Park (2017, pp. 30–31), the possibility statement that H "may be the best of a bad lot" (van Fraassen, 1989, p. 143). Similarly, the premise of the argument from a brute fact is a possibility statement that a fact might be brute.

Why do I fuss about the fact that the premises are possibility statements? When users of IBE claim that H is true, they do not mean that it is guaranteed to be true but rather that it is likely to be true. Consequently, their claim is compatible with the possibility that H might be the best of a bad lot. Similarly, when users of IBE attempt to explain a fact, they do not believe that it is guaranteed to be non-brute but rather that it is likely to be non-brute. Accordingly, their attempt to explain it is compatible with the possibility that it might be brute. In short, the two possibilities do not pose any threat to the use of IBE.

In order to pose a threat to the claim that H is likely to be true, van Fraassen has to establish not the possibility statement that H might be the best of a bad lot, but rather the probabilistic statement that H is likely to be the best of a bad lot. If it is likely that H is the best of a bad lot, then users of IBE would not be justified in believing H. Unfortunately, van Fraassen does not attempt to establish that H is likely to be the best of a bad lot; he only claims that H might be the best of a bad lot.

The same holds for de Ray. He has to establish not the possibility statement that a fact might be brute, but rather the probabilistic statement that it is likely to be brute. For example, he should not establish that we "can imagine that nothing at all explains the appearance of pellets on the floor and strange marks on the cheese this morning" (de Ray, 2022, p. 382), but rather that it is likely that nothing explains the phenomenon. If a fact is likely to be brute, users of IBE would not be justified in believing that it has an explanation. Unfortunately, de Ray does not attempt to establish that a fact is likely to be brute; he only claims that we can imagine that it is brute.

4.4.3 Confounding a Means and an End

In this subsection, I argue that both the arguments from a bad lot and a brute fact result from confounding a means and an end.

Hume confuses a means and an end when he claims that the uniformity principle is required for inductions to be justified (Sect. 1.3.2 of Chap. 1). The unavoidable consequence of confusing the means and the end is skepticism about both the means and the end, i.e., about both induction and the uniformity principle. To use an analogy, you confuse a means and an end when you claim that to be on the roof is required to use a ladder. The unavoidable consequence of confusing the means and the end is that you can neither use the ladder nor be on the roof.

The same objection applies mutatis mutandis to the argument from a bad lot. Suppose that you and I investigate the same phenomena, but that you and I hold different hypotheses. Before you hold your hypothesis, you eliminated one hundred false hypotheses. By contrast, before I hold my hypothesis, I eliminated only one false hypothesis. In addition, you do not know which hypothesis I eliminated, and I do not know either which hypotheses you eliminated. Under such circumstances, your hypothesis is more likely to have been chosen from a good lot than is my hypothesis, and you are more entitled than I am to believe that the true hypothesis is conceived than unconceived. It follows that using IBE is a means to arrive at the privilege principle rather than the other way around.

As a result of repeatedly using IBE, we believe that a true hypothesis is more likely to be found in the collection of the rivals that we considered than in the collection of the rivals that we did not. It is not the case that we repeatedly use IBE as a result of believing that a true hypothesis is more likely to be found in the collection of the rivals that we considered than in the collection of the rivals that we did not. Therefore, the use of IBE supports the privilege principle rather than the other way around, and it is to confuse the means and the end for van Fraassen to say that the privilege principle is required for IBE to be justified. The inevitable consequence of confounding the means and the end is skepticism about both IBE and the privilege principle.

There is a general consensus between scientific realists and antirealists that current theories such as the kinetic theory are more likely to be true than past theories such as the caloric theory, although there is no general consensus regarding the issue of whether current theories are successful enough to be true (Müller, 2015, p. 406; Alai, 2017, p. 3283). The general consensus implies that current theories are more likely than past theories to have been chosen from good lots. Why is it that the set of the rivals that current scientists considered is more likely to include true theories than the set of the rivals that past scientists considered? My answer is that scientists have been repeatedly using IBE in their investigations of the world. As a result, current scientists are more likely than past scientists to have hit on the ranges of theories that include true theories. This consideration of the difference between past and present scientists reinforces my view that IBE supports the privilege principle rather than the other way around.

The same objection applies mutatis mutandis to the argument from a brute fact. The use of IBE is a means to arrive at the explainability principle rather than the other way around. As a result of repeatedly using IBE, we believe that the world is explainable. It is not the case that we repeatedly use IBE as a result of believing that the world is explainable (Park, 2023, pp. 478–479). For example, as a result of using IBE regarding the denial of a promotion, the fall of a rock, the existence of gravitational wave, and so on, we are entitled to believe the explainability principle. It is not the case that as a result of being entitled to believe the explainability principle, we use IBE regarding the denial of a promotion, the fall of a rock, the existence of gravitational wave, and so forth. Therefore, the use of IBE is required for the explainability principle to be justified rather than the other way around, and it is to confuse the means and the end for de Ray to say that the explainability principle is required for IBE to be justified. The unavoidable consequence of confounding the means and the end is skepticism about both IBE and the explainability principle.

There is also a general consensus between scientific realists and antirealists that current theories such as the kinetic theory explain more phenomena than past theories such as the caloric theory (Psillos, 2018). Moreover, current scientists have far more scientific theories than past scientists (Fahrbach, 2011a, 2011b; Park, 2011; Mizrahi, 2013). Consequently, current scientists can explain far more phenomena than past scientists, and current scientists are more entitled than past scientists to believe that the world is explainable. Why is it that current scientists are more entitled to believe the explainability principle than past scientists? My answer is that scientists have been repeatedly using IBE in their investigations of the world. In the course of using IBE in diverse contexts, they have encountered more and more phenomena that had explanations. As a result, they became more entitled to believe that the world is explainable. This consideration of the difference between current and past scientists bolsters my view that IBE supports the explainability principle rather than the other way around.

It is due to the false presupposition that IBE requires the privilege and explainability principles that the arguments from a bad lot and a brute fact impose excessive requirements on the use of IBE. The argument from a bad lot imposes the excessive requirement that before using IBE, users of IBE should justify the privilege principle without using IBE. The argument from a brute fact imposes the excessive requirement that before using IBE, users of IBE should justify the explainability principle without using IBE. As a result, no one can meet the requirements, and no one can justifiably use IBE and believe that H is true. Even van Fraassen and de Ray themselves cannot meet the excessive requirements, and they cannot justifiably believe their own philosophical views, as we have seen in Sect. 4.3 above.

Just as a ladder is a means and to be on the roof is an end, so IBE is a means and the privilege and explainability principles are ends. However, the arguments from a bad lot and a brute fact erroneously presuppose that the privilege and explainability principles are the means and the use of IBE is the end. Just as to say that the use of a ladder requires to be on the roof is to invite the situation where no one can use a ladder and be on the roof, so to say that the use of IBE requires the privilege and explainability principles is to invite the situation where no one can use IBE and arrive at the privilege and explainability principles.

4.5 Conclusion

The use of IBE entitles us to believe the privilege and explainability principles rather than the other way around. However, the arguments from a bad lot and a brute fact erroneously presuppose that the privilege and explainability principles entitle us to use IBE, thereby confusing the means and the ends. The cost of confusing the means and the ends is the excessive requirements on the use of IBE, skepticism about our best scientific theories, and the backfires on van Fraassen's use and de Ray's use of IBE. This chapter can be summed up as follows: "Confusing a means and an end is a shortcut to skepticism."

References

Alai, M. (2017). Resisting the historical objections to realism: Is Doppelt's a viable solution? *Synthese, 194*(9), 3267–3290.

de Ray, C. (2022). A new epistemological case for theism. *Religious Studies, 58*(2), 379–400.

Fahrbach, L. (2011a). How the growth of science ends theory change. *Synthese, 180*(2), 139–155.

Fahrbach, L. (2011b). Theory change and degrees of success. *Philosophy of Science, 78*(5), 1283–1292.

Hempel, C. (1966). *Philosophy of natural science*. Prentice-Hall.

Mizrahi, M. (2013). The pessimistic induction: A bad argument gone too far. *Synthese, 190*(15), 3209–3226.

Mizrahi, M. (2020). *The relativity of theory: Key positions and arguments in the contemporary scientific realism/antirealism debate*. Springer.

Müller, F. (2015). The pessimistic meta-induction: Obsolete through scientific progress? *International Studies in the Philosophy of Science, 29*(4), 393–412.

Park, S. (2011). A confutation of the pessimistic induction. *Journal for General Philosophy of Science, 42*(1), 75–84.

Park, S. (2017). Scientific antirealists have set fire to their own houses. *Prolegomena, 16*(1), 23–37.

Park, S. (2019). The disastrous implications of the 'English' view of rationality in a social world. *Social Epistemology, 33*(1), 88–99.

Park, S. (2023). Inference to the best explanation, naturalism, and theism. *Implicit Religion, 24*, 472–485.

Psillos, S. (2018). Realism and theory change in science. In E. N. Zalta (Ed.), *The Stanford encyclopedia of philosophy*. Retrieved from https://plato.stanford.edu/archives/sum2018/entries/realism-theory-change/

van Fraassen, B. (1980). *The scientific image*. Oxford University Press.

van Fraassen, B. (1989). *Laws and symmetry*. Oxford University Press.

Weintraub, R. (2013). Induction and inference to the best explanation. *Philosophical Studies, 166*(1), 203–216.

Wray, K. (2021). Induction, rationality, and the realism/antirealism debate: A reply to Shech. *International Studies in the Philosophy of Science, 34*(4), 243–247.

Chapter 5
The Template for Generating Philosophical Problems

Abstract A template for generating philosophical problems consists of the following three steps. First, skeptics argue that some particular beliefs might be false. Second, they suggest that a general belief is required for some particular beliefs to be justified. Third, they claim that no particular belief is justified, and that those who hold some particular beliefs have the burden of justifying a general belief without relying on a particular belief. This template can be used to generate philosophical problems, such as the problem of induction, the argument from a bad lot, the argument from a brute fact, and other philosophical problems. They all result from confusing a means and an end.

Keywords Bad lot · Brute fact · Induction · Inference to the best explanation · Template

5.1 Introduction

I discussed David Hume's (1993, pp. 15–25) problem of induction in Chaps. 1–3. The problem of induction generates skepticism about inductive inferences, including inference to the best explanation (IBE). In Chap. 4, I discussed the argument from a bad lot (van Fraassen, 1989, p. 143) and the argument from a brute fact (Weintraub, 2013, p. 213; de Ray, 2022). These two arguments generate skepticism about IBE. This chapter aims to provide readers with a template for generating philosophical problems, including the aforementioned three problems, thereby shedding new light on them.

The present chapter is structured as follows. In Sect. 5.2, I point out two similarities among the problem of induction, the argument from a bad lot, and the argument from a brute fact. In Sect. 5.3, I extract a template for generating philosophical problems from the two similarities, and then use the template to generate several other philosophical problems. In Sect. 5.4, I offer four rules on how to generate sophisticated philosophical problems with the use of the template. In Sect. 5.5, I argue that the problem of induction complies with the three of the four rules, and

S. Park, *Induction, Science, and Morality*, Synthese Library 520,
https://doi.org/10.1007/978-3-032-00032-3_5

thus that it attracted a lot of attention from philosophers for long in the history of philosophy. In Sect. 5.6, I argue that the template cannot be used to generate Cartesian, mathematical, and moral skepticism.

This chapter is intended to be useful to those who aim to gain a new insight into the problem of induction, the argument from a bad lot, the argument from a brute fact, and other philosophical problems generated with the use of the template.

5.2 The Two Similarities

In this section, I point out two similarities among the problem of induction, the argument from a bad lot, and the argument from a brute fact.

Hume claims that the faculty of reason does not justify the conclusions of inductions because the negations of the conclusions do not involve contradictions (Sect. 1.2 of Chap. 1). This argument implies that the faculty of reason does not justify the conclusions of inductions ultimately because it is possible that the conclusions of inductions might be false. Hume also claims that experience does not justify the conclusions of inductions because there is no necessary connection between the properties that the inductions assert are connected with each other (Sect. 1.2 of Chap. 1). This argument implies that experience does not justify the conclusions of inductions ultimately because it is possible that the conclusions of inductions might be false. Note that these two arguments are built upon the possibility that the conclusions of inductions might be false.

It is much more difficult to establish that the conclusions of inductions are *likely* to be false than to establish that they might *possibly* be false. Only imaginative power is required to establish the possibility, while an empirical study is required to establish the high probability. In the empirical study, skeptics should randomly select some conclusions of inductions from the population of all conclusions of inductions and then show that the selected conclusions are mostly false. Hume does not use the random sampling method to develop the problem of induction. He only entertains the possibility that the conclusions of inductions might be false.

Bas van Fraassen's argument from a bad lot consists of the premise that the conclusions of IBE might be from bad lots, which implies that they might be false, and the conclusion that a general belief is required for the conclusions of IBE to be justified. In Sect. 4.2 of Chap. 4, the general belief is named as "the privilege principle," according to which the conclusions of IBE are more likely to be chosen from good lots than from bad lots. Users of IBE are required to justify the privilege principle before they use IBE. Van Fraassen does not use the random sampling method to establish skepticism about IBE. Specifically, he does not randomly select some conclusions of IBE from the population of all conclusions of IBE and then show that the selected conclusions are mostly from bad lots. He only entertains the possibility that the conclusions of IBE might be from bad lots, which implies that they might be false.

Christophe de Ray's argument from a brute fact consists of the premise that a fact might not have an explanation, i.e., that it might be brute, and the conclusion that a general belief is required for the use of IBE to be justified. In Sect. 4.2 of Chap. 4, the general belief is named as "the explainability principle," according to which the world is explainable. Users of IBE are required to justify the explainability principle before they use IBE. De Ray does not use the random sampling method to establish skepticism about IBE. Specifically, he does not randomly select some facts from the population of all facts in the world and then show that the selected facts are mostly brute. He only entertains the possibility that a fact might be brute, which implies that the conclusions of IBE might be false.

There is a similarity among the problem of induction, the argument from a bad lot, and the argument from a brute fact: they all are grounded in the mere possibility that some particular beliefs might be false. Specifically, the problem of induction trades on the mere possibility that the conclusions of inductions might be false. The argument from a bad lot relies on the mere possibility that the conclusions of IBE might be from bad lots, which implies that they might be false. The argument from a brute fact rests on the mere possibility that a fact might be brute, i.e., it might be false that the fact has an explanation. In short, all the three arguments are grounded in mere possibilities rather than in high probabilities.

There is another similarity among the three arguments: they all suggest that a general belief is required for some particular beliefs to be justified. Hume's problem of induction claims that the uniformity principle is required for (the conclusions of) inductions to be justified. Van Fraassen's argument from a bad lot asserts that the privilege principle is required for (the conclusions of) IBE to be justified. De Ray's argument from a brute fact maintains that the explainability principle is required for (the conclusions of) IBE to be justified.

All the three arguments confuse a means and an end, i.e., they confuse the direction of requirement between some particular beliefs and a general belief, taking a general belief to be a required means to arrive at some particular beliefs, when in fact some particular beliefs are a required means to arrive at a general belief. In other words, they erroneously presuppose that a general belief is required for some particular beliefs to be justified when in fact some particular beliefs are required for a general belief to be justified. As a result, they invite the situation where no one can justifiably hold both some particular beliefs and a general belief, i.e., they generate skepticism about both some particular beliefs and a general belief.

5.3 The Template

In Sect. 5.2 above, I exposed the two similarities among the problem of induction, the argument from a bad lot, and the argument from a brute fact. In this section, I infer a template from the two similarities and then use it to generate several other philosophical problems.

The template consists of the following three steps. First, you point out that some particular beliefs might be false. It requires imaginative work rather than empirical work to show that they might be false. Second, you suggest that a general belief is required for them to be justified. Third, you claim that they are not justified, and that those who hold them have the burden of justifying a general belief without relying on one of them. You might attempt to justify a general belief yourself with the appeal to one of them, thereby demonstrating to your readers that it is circular to do so, and that it is a challenging task to justify a general belief. Going through these three steps results in a philosophical problem that might be of interest to your philosophical colleagues.

The problem of induction fits this template. First, Hume observes that the conclusions of inductions might be false. Second, he claims that the uniformity principle is required for inductions to be justified. Third, he claims that inductions are "entirely arbitrary," i.e., that no induction is justified (Hume, 1993, p. 18), and that inductivists, those who endorse inductions, have the burden of justifying the uniformity principle without relying on an induction. Going through these three steps results in the problem of induction.

The argument from a bad lot also fits the template. First, van Fraassen observes that the conclusions of IBE might be from bad lots, which implies that they might be false. Second, he suggests that the privilege principle is required for IBE to be justified. Third, he claims that IBE is not justified, and that users of IBE have the burden of justifying the privilege principle without using IBE. Going through these three steps results in the argument from a bad lot.

The argument from a brute fact also fits the template. First, de Ray observes that a fact might not have an explanation, which implies that the conclusions of IBE might be false. Second, he suggests that the explainability principle is required for IBE to be justified. Third, he claims that IBE is not justified, and that users of IBE have the burden of justifying the explainability principle without using IBE. Going through these three steps results in the argument from a brute fact.

Any philosophical problem whose production process fits the template above is a product of confusing a means and an end. It is to confound a means and an end for Hume to say that the uniformity principle is required for inductions to be justified, for van Fraassen to say that the privilege principle is required for IBE to be justified, and for de Ray to say that the explainability principle is required for IBE to be justified. It is not surprising that all these three philosophical problems generate skepticism about both inductions and the three principles.

So far in this section, I stated the template, expounded how the process of producing the three philosophical problems fits the template, and then claimed that they all involve confusing a means and an end. From now on, I use the template to generate several other philosophical problems, some of which are already familiar to philosophers.

The template can be used to create a philosophical problem that I call "the problem of memory." First, skeptics point out that memory is fallible, i.e., that memorial beliefs might be false. Second, they suggest that a general belief is required for memorial beliefs to be justified. I call the general belief "the memorial principle," according to which memory is reliable, i.e., we tend to remember past events correctly. Third, skeptics claim that no memorial belief is justified, and that those who hold memorial beliefs have the burden of justifying the memorial principle without relying on a memorial belief. Going through these three steps leads to the problem of memory.

The template can be used to produce a philosophical problem that I earlier called "the problem of mind" in Sect. 1.3.5 of Chap. 1. First, skeptics point out that the mind is fallible, i.e., that beliefs might be false. Second, they suggest that a general belief is required for beliefs to be justified. I call the general belief "the reliability principle," according to which the mind is reliable. Third, skeptics claim that no belief is justified, and that those who hold some beliefs have the burden of justifying the reliability principle without appealing to a belief. Going through these three steps produces the problem of mind.

The template can be used to create a philosophical problem that I earlier called "the problem of inference to the best competitor (IBC)" in Sect. 2.8 of Chap. 2. IBC is used when coherentists (Lehrer, 2000, p. 170) conclude that a belief is justified on the grounds that it defeats its competitors. First, skeptics point out that even if a belief defeats its competitor, it might be false. Second, they suggest that a general belief is required for IBC to be justified. I earlier called the general belief "the nonexistence principle," according to which it is more likely than not that unconceived competitors that would defeat the best competitor do not exist. Third, skeptics claim that IBC is not justified, and that users of IBC have the burden of justifying the nonexistence principle without relying on IBC. Going through these three steps creates the problem of IBC.

The template can give rise to a philosophical problem that I call "the problem of simplicity." First, scientific antirealists point out that simple scientific theories might be false while complex ones might be true. Second, they suggest that a general belief is required for simple theories to be justified. I call the general belief "the simplicity principle," according to which the world is simple. Third, scientific antirealists claim that no simple theory is justified, and that scientific realists have the burden of justifying the simplicity principle without relying on a simple theory. Going through these three steps gives rise to the problem of simplicity.

The template can yield a philosophical problem that I call "the problem of appearance." First, scientific antirealists point out that some scientific theories might be false although they appear to be true, i.e., although they are successful (Park, 2021). For example, past theories, such as the Ptolemaic theory, appeared to be true but they were false. Second, scientific antirealists suggest that a general belief is required for successful theories to be justified. I call the general belief "the transparency principle," according to which the appearance of successful theories tends to accord with their reality, i.e., successful theories tend to be true. Third, scientific antirealists claim that no successful theory is justified, and that scientific realists, those who believe successful theories, have the burden of justifying the transparency principle without invoking a successful theory. Going through these three steps yield the problem of appearance.

The template can be used to create a philosophical problem that I call "the problem of generalization." First, skeptics point out that general beliefs, such as the belief that all crows are black and the belief that all gold bars are yellow, might be false. Second, skeptics suggest that an overarching general belief is required for general beliefs to be justified. I call the overarching general belief "the generalizability principle," according to which the world is generalizable. Third, skeptics claim that no general belief is justified, and that those who hold general beliefs have the burden of justifying the generalizability principle without using a general belief. Going through these three steps creates the problem of generalization.

Finally, the template can generate a philosophical problem that I call "the problem of sapphires." First, skeptics point out that some particular beliefs about the color of sapphires, such as the belief that this sapphire is blue and the belief that that sapphire is blue, might be false. Second, skeptics suggest that a general belief is required for some particular beliefs about the color of sapphires to be justified. I call the general belief "the sapphire principle," according to which all sapphires are blue. Third, skeptics claim that no particular belief about the color of sapphires is justified, and that those who hold some particular beliefs about the color of sapphires have the burden of justifying the sapphire principle without relying on one of them. Going through these three steps generates the problem of sapphires.

In this section, I generated several philosophical problems with the use of the template. The last one, the problem of sapphires, differs from the rest of them. The problem of sapphires would attract no philosopher's attention while the other philosophical problems would attract many philosophers' attention. What accounts for the difference between them? I answer this question in the next section.

5.4 How to Generate Sophisticated Philosophical Problems

Some philosophical problems generated with the use of the template are of high quality, while others are of low quality. How can we generate sophisticated philosophical problems with the use of the template? In this section, I develop the answer that the more we observed the following four rules, the more sophisticated the resulting philosophical problem will be.

(i) A general belief should sound profound. It does not sound profound that all sapphires are blue, but it sounds profound that nature is uniform, that the world is explainable, that the world is simple, and that the world is generalizable. If a general belief sounds profound, some philosophers will think that the justification for a profound truth about the world is required to solve the philosophical problem generated with the use of the template. As a result, they will be motivated to solve it in vain.

(ii) A general belief should be highly general. It is not a highly general belief that all sapphires are blue, but it is highly general beliefs that nature is uniform, that the world is explainable, that the world is simple, and that the world is generalizable. If a belief is highly general, some philosophers will think that it is innate. No philosopher thinks that we are born with the general belief that all sapphires are blue, but some philosophers will think that we are born with the general beliefs that nature is uniform, that the world is explainable, that the world is simple, and that the world is generalizable.

Some philosophers will say that the justification for highly general beliefs comes from the faculty of reason, i.e., that they are synthetic a priori knowledge. As a result, philosophers will be divided into two camps and have a debate over whether they are synthetic a priori knowledge or synthetic a posteriori knowledge, instead of unearthing the truth that the philosophical problems that they were trying to solve are merely the product of confusing a means and an end.

(iii) An alternative to a general belief should be formulated. For example, when you generate the problem of simplicity, you should formulate what might be called "the complexity principle," according to which the world is complex, and then challenge believers of simple theories to present reasons for preferring the simplicity principle to the complexity principle. Your challenge will trigger a debate between rival philosophers over which principle is true, i.e., over whether the world is simple or complex. The debaters on both sides will assume that once the world is shown to be simple, the problem of simplicity will be solved, applying the outcome of their debate to the problem of simplicity in vain.

Meanwhile, the debaters will pay no attention to the issue of whether the simplicity principle is required for simple theories to be justified or vice versa. As a result, they will stay away, unbeknownst to themselves, from the truth that it is merely to confuse a means and an end to say that the simplicity principle is required for simple theories to be justified, i.e., from the truth that the problem of simplicity is merely a product of confusing the direction of requirement between the simplicity principle and simple theories. In short, an alternative to a general belief will serve as a distracting device.

(iv) It should be somewhat opaque whether a general belief is required for some particular beliefs to be justified or vice versa. If it is transparent that some particular beliefs are required for a general belief to be justified, it will be obvious that a philosophical problem generated with the use of the template is a product of confusing a means and an end. As a result, philosophers will dismiss it immediately.

Consider the supporting relationship between some beliefs about the color of sapphires and the general belief that all sapphires are blue. It is obvious that some particular beliefs about the color of sapphires are required for the general belief to be justified rather than the other way around. As a result, no mental energy is required for philosophers to see that the problem of sapphires is merely a product of confusing the direction of requirement between some particular beliefs and a general belief. Such a philosophical problem is fated to attract no philosopher' attention.

How about the supporting relationship between the privilege principle and IBE? It is somewhat opaque whether the privilege principle is required for IBE to be justified or vice versa. As a result, a fair amount of mental energy is required to see that the argument from a bad lot is merely a product of confusing the direction of requirement between IBE and the privilege principle (Sect. 4.4.3 of Chap. 4). Such a philosophical problem is sure to receive a lot of attention from philosophers and last long in the history of philosophy.

The same holds for the supporting relationships between the uniformity principle and inductions, between the explainability principle and IBE, between the simplicity principle and simple theories, and between the generalizability principle and general beliefs. It is somewhat opaque whether general beliefs are required for some particular beliefs to be justified or vice versa in such cases. For this reason, some philosophers might even think that the general beliefs are innate, that they serve as lenses through which we see the world, that they are justified a priori, and that they justify some particular beliefs.

5.5 The Problem of Induction

In this section, I argue that the problem of induction is a sophisticated philosophical problem because it complies with the three of the four rules stated in Sect. 5.4 above.

The problem of induction observes the first and second rules. The uniformity principle sounds profound. As a result, it appears that the justification for the profound truth about the world is required to solve the problem of induction. The uniformity principle is highly general in that it concerns every corner of the world. As a result, it appears that it is justified a priori and justifies inductions.

The problem of induction does not observe the third rule that an alternative to a general belief should be formulated, given that Hume did not formulate an alternative to the uniformity principle. However, I (2017, 2018, 2022, pp. 76–79) formulated an alternative to it that I call "the disuniformity principle" according to which nature is disuniform. The disuniformity principle is to counter-induction as the uniformity principle is to induction (Sect. 1.4 of Chap.1).

What would be a possible effect of formulating the disuniformity principle? Inductive skeptics might challenge inductivists to present reasons for preferring the uniformity principle to the disuniformity principle. Their challenge will trigger a debate between rival philosophers over which principle is true, i.e., over whether nature is uniform or disuniform. The debaters on both sides will assume that once nature is shown to be uniform, the problem of induction will be solved, applying the outcome of their debate to the problem of induction in vain.

Meanwhile, the debaters will pay no attention to the issue of whether the uniformity principle is a means to arrive at inductions or vice versa, and to the issue of whether the disuniformity principle is a means to arrive at counter-inductions or vice versa. As a result, they will stay away, unbeknownst to themselves, from the truth that it is merely to confuse a means and an end to say that the uniformity principle is required for inductions to be justified, and from the truth that the disuniformity principle is required for counter-inductions to be justified. In short, the disuniformity principle will serve as a distracting device.

The problem of induction follows the fourth rule that it should be somewhat opaque whether a general belief is required for some particular beliefs to be justified or vice versa. Consider the supporting relationship between the uniformity principle and inductions. It is clear that inductions support the uniformity principle. After all, we infer the uniformity principle from such inductions as (S) and (N):

(S)
Observed sapphires are blue.
∴ Unobserved sapphires are blue.

(N)
Neptune was observed to revolve around the sun in the past.
∴ Neptune will revolve around the sun in the future.

We first form such inductions and then come to believe the uniformity principle. Without first forming the inductions, we would not come to believe the uniformity principle. Consequently, inductions support the uniformity principle.

Suppose now that after you have embraced the uniformity principle on the basis of some inductions, you embraced the following new induction, (U):

(U)
Uranus was observed to revolve around the sun in the past.
∴ Uranus will revolve around the sun in the future.

What supports the conclusion of (U)? Obviously, the premise of (U) does. But what about the uniformity principle? The uniformity principle was in your mind before you formed (U). So the uniformity principle seems to support the conclusion of (U) to some degree. As a result, it is somewhat opaque whether inductions support the uniformity principle or vice versa, and hence which is a means and which is an end. It is not surprising that the problem of induction has persisted for hundreds of years in the history of philosophy.

What would be the reason for thinking, then, that the problem of induction is nonetheless a product of confusing a means and an end? It makes the strong claim that the uniformity principle is required for inductions to be justified, i.e., that inductions would not be justified without the uniformity principle. Compare the support that the uniformity principle provides for the conclusion of (U) with the support that the premise of (U) provides for it. The latter is required, while the former is optional, for the conclusion of (U) to be justified.

Moreover, inductions are required for the uniformity principle to be justified, i.e., the uniformity principle would not be justified without inductions. It is wrong to think that the uniformity principle is innate, that it is justified a priori, and then that it justifies inductions. Therefore, it is to confuse the direction of requirement between inductions and the uniformity principle to say that the uniformity principle is required for inductions to be justified.

To use an analogy, suppose that you observe some sapphires, which are all blue, and then form the general belief that all sapphires are blue. You observe a new sapphire, and then form the new belief that it is blue. The support for this new belief comes from the current observation of the new sapphire, from the previous observation of other sapphires, and from the general belief that all sapphires are blue. Compare the support that the general belief provides for the new belief with the support that the current and previous observations provides for it. The latter is required, while the former is optional, for the new belief to be justified.

Moreover, some particular beliefs about the color of sapphires are required for the general belief to be justified, i.e., the general belief would not be justified at all without the particular beliefs. It is wrong to think that the general belief is innate, that it is justified a priori, and then that it justifies the particular beliefs. Therefore, it is to confuse the direction of requirement between the particular beliefs and the general belief to say that the general belief is required for the particular beliefs to be justified. The same holds for Hume's contention that the uniformity principle is required for inductions to be justified.

To recap, the uniformity principle seems to support some inductions to some degree. It is due to this feature of the uniformity principle that philosophers are misled into thinking that the uniformity principle is required for inductions to be justified, and that it is difficult to tell whether the uniformity principle is required for inductions to be justified or vice versa. However, the problem of induction is nevertheless a product of confusing a means and an end. The support that some inductions provide for the uniformity principle is required for it to be justified while the support that the uniformity principle provides for other inductions is optional for them to be justified.

5.6 Cartesian, Mathematical, and Moral Skepticism

Cartesian, mathematical, and moral skepticism are views, respectively, that we do not have enough evidence for the existence of a physical fact, a mathematical fact, and a moral fact. Can the template be used to generate these views? Initially, it appears that it can, but on close examination, it cannot. So I argue in this section.

Cartesian skeptics might be tempted to use the template as follows. First, they argue that physical beliefs (i.e., beliefs about physical objects) might be false, given that an evil demon rather than the world might cause experience in our minds. Second, they suggest that a general belief is required for physical beliefs to be justified. It might be called "the causation principle," according to which the world causes experience in our minds. Third, they claim that no physical belief is justified, and that those who hold physical beliefs have the burden of justifying the causation principle without invoking a physical belief. Going through these three steps results in Cartesian skepticism.

It appears that the process of generating Cartesian skepticism fits the template. On close examination, however, it does not. The template only fits the cases in which skeptics claim that a general belief is required for some particular beliefs to be justified. However, Cartesian skeptics do not make such a claim; they rather claim that we do not have good evidence for the existence of physical facts. Once the existence of a single physical fact is established, they would readily infer that there are other physical facts. For example, once it is established that some crows are black, they would happily infer that other crows are also black. They do not have a problem with the inference from some particular physical beliefs to a general physical belief, but rather with the inference from a particular physical belief to its corresponding particular physical fact.

Mathematical skeptics might be tempted to use the template as follows. First, they argue that mathematical beliefs, such as the belief that $4 + 3 = 7$, might be false, given that mathematical facts are imperceptible and we can imagine that they do not exist. Any statement that is made true by no fact is false. Second, mathematical skeptics suggest that a general belief is required for mathematical beliefs to be justified. It might be called "the mathematical realist principle," according to which there are mathematical facts in the abstract world to which some mathematical beliefs correspond. The abstract world is the world in which there are abstract facts, including mathematical ones. It contrasts with the concrete world in which there are concrete facts, such as the fact that water is H_2O. Third, mathematical skeptics claim that no mathematical belief is justified, and that holders of mathematical beliefs have the burden of justifying the mathematical realist principle without relying on a mathematical belief. Going through these three steps result in mathematical skepticism.

It appears that the process of generating mathematical skepticism fits the template. On close examination, however, it does not. Again, the template only fits the cases in which skeptics claim that a general belief is required for some particular beliefs to be justified. However, mathematical skeptics do not make such a claim; they rather claim that we do not have good evidence for the existence of mathematical facts. Once the existence of a single mathematical fact is established, mathematical skeptics would readily infer that there are other mathematical facts. For example, once it is established that the sum of the four angles of some triangles is 360°, they would happily infer that the sum of the four angles of other triangles is also 360°. They do not have a problem with the inference from some particular mathematical beliefs to a general mathematical belief, but rather with the inference from a particular mathematical belief to its corresponding particular mathematical fact.

Moral skeptics might be tempted to use the template as follows. First, they argue that moral beliefs, such as the belief that murder is (morally) wrong and the belief that blood donation is right, might be false, given that moral facts are imperceptible and we can imagine that they do not exist. Any statement that is made true by no fact is false. Second, moral skeptics suggest that a general belief is required for some moral beliefs to be justified. I call the general belief "the moral realist principle," according to which some moral facts exist. Third, moral skeptics claim that no moral belief is justified, and that holders of moral beliefs have the burden of justifying the moral realist principle without relying on a moral belief. Going through these three steps result in moral skepticism.

It appears that the process of generating moral skepticism fits the template. On close examination, however, it does not. Again, the template only fits the cases in which skeptics claim that a general belief is required for some particular beliefs to be justified. However, moral skeptics do not make such a claim; they rather make the claim that we do not have good evidence for the existence of moral facts. Once the existence of a single moral fact is established, moral skeptics would readily infer that there are other moral facts. For example, once it is established that stealing was wrong, they would happily infer that stealing will be wrong. They do not have a problem with the inference from some particular moral beliefs to a general moral belief, but rather with the inference from a particular moral belief to its corresponding particular moral fact.

Where does all this lead us? Cartesian, mathematical, and moral skeptics do not confuse a means and an end when they advance skepticism about physical, mathematical, and moral facts. None of them erroneously presupposes that the causation, mathematical, and moral realist principles are required for some particular physical, mathematical, and moral beliefs to be justified, respectively. In addition, there can be legitimate and interesting debates between realists and antirealists over whether physical, abstract, and moral facts exist.

I jump into the debate between moral realists and antirealists, and into the debate between mathematical realists and antirealists in Chaps. 10–15. Readers will see that I do not rely on the template to criticize mathematical and moral realism, and thus that I do not confuse a means and an end.

5.7 Conclusion

The template for generating philosophical problems consists of the following three steps: (a) arguing that some particular beliefs might be false, (b) suggesting that a general belief is required for some particular beliefs to be justified, (c) and then claiming that no particular belief is justified, and that holders of some particular beliefs have the burden of justifying the general belief without relying on one of them.

The more you abide by the following four rules, the more sophisticated your philosophical problem will be. (1) A general belief should sound profound. (2) A general belief should be highly general. (3) An alternative to a general belief should be formulated. (4) It should be somewhat opaque whether a general belief is required for some particular beliefs to be justified or vice versa.

The problem of induction is a sophisticated philosophical problem because it observes the three of the four rules on how to use the template. The uniformity principle sounds profound and is highly general. It is somewhat opaque whether inductions are required for the uniformity principle to be justified or vice versa. It is not surprising that the problem of induction attracted a lot of attention from philosophers for long in the history of philosophy.

Cartesian, mathematical, and moral skepticism cannot be generated with the use of the template. The template generates skepticism about an inference from some particular beliefs to a general belief. However, Cartesian, mathematical, and moral skepticism do not dispute the inference from some particular beliefs to a general belief, but rather the existence of a physical fact, a mathematical fact, and a moral fact.

This chapter can be summed up as follows: "The template harbors the confusion of a means and an end."

References

de Ray, C. (2022). A new epistemological case for theism. *Religious Studies, 58*(2), 379–400.

Hume, D. (1993). *An enquiry concerning human understanding*. Hackett Publishing.

Lehrer, K. (2000). *Theory of knowledge*. Westview Press.

Park, S. (2017). The uniformity principle vs. the disuniformity principle. *Acta Analytica, 32*(2), 213–222.

Park, S. (2018). The anti-induction for scientific realism. *Grazer Philosophische Studien, 95*(3), 329–342.

Park, S. (2021). On the argument from double spaces: A Reply to Moti Mizrahi. *Social Epistemology Review and Reply Collective, 10*(2), 1–6.

Park, S. (2022). *Embracing scientific realism*. Springer.

van Fraassen, B. (1989). *Laws and symmetry*. Oxford University Press.

Weintraub, R. (2013). Induction and inference to the best explanation. *Philosophical Studies, 166*(1), 203–216.

Chapter 6
Inductive Skepticism and the Scientific Realism Debate

Abstract If scientific antirealists retreated to inductive skepticism, they would have to give up their positive philosophical views, such as constructive empiricism, the selectionist explanation, the contextual theory, and the permissive view of rationality. They would also have to give up their negative arguments against scientific realism, such as the pessimistic induction, the argument from a bad lot, and the argument from indifference. The local theory of induction generates inductive skepticism and thus undermines both scientific realism and antirealism.

Keywords Hume · Inductive skepticism · Local theory · Problem of induction · Scientific antirealism

6.1 Introduction

This chapter explores the relationship between inductive skepticism and the scientific realism debate. I answer the following questions concerning the relationship between them. How does scientific antirealism relate to inductive skepticism? What are the costs that scientific antirealists would have to pay if they retreated to inductive skepticism? How does the local theory of induction (Norton, 2003) relate to scientific realism and antirealism? Does it help scientific realism? Should localists, those who espouse the local theory, reject scientific antirealism and accept scientific realism? I provide my answers to these questions, and they would be disagreeable to Bas van Fraassen (2007), K. Brad Wray (2021), and Elay Shech (2021).

This chapter is organized as follows. In Sect. 6.2, I distinguish between inductive skepticism and scientific antirealism. In Sects. 6.3 and 6.4, I argue that scientific antirealists' retreat to inductive skepticism involves the cost of losing their positive philosophical views, such as constructive empiricism, the selectionist explanation, the contextual theory, and the permissive view of rationality, and the cost of losing their negative arguments against scientific realism, such as the pessimistic induction, the argument from a bad lot, and the argument from indifference. In Sect. 6.5,

S. Park, *Induction, Science, and Morality*, Synthese Library 520,
https://doi.org/10.1007/978-3-032-00032-3_6

I argue that the local theory conflicts with both scientific realism and antirealism, contrary to what Shech (2021) thinks.

The present chapter will be rewarding to those who aim to know what costs are involved in scientific antirealists' retreat to inductive skepticism, and to those who wonder whether the local theory helps or hurts scientific realism and antirealism.

6.2 Inductive Skepticism and Scientific Antirealism

In this section, I define scientific antirealism and then explain how it differs from inductive skepticism.

There are diverse forms of scientific antirealism in the literature. One of them asserts that what a scientific theory, T, says about observables is (mostly) true, but it is silent about what T says about unobservables. Antirealists advocating this variant of antirealism observe that T makes true predictions, and then infer that it will make true predictions. They believe that it is empirically adequate or approximately so. Let me call this variant of scientific antirealism "antirealism" from now on. To use an analogy, antirealists observe some yellow gold bars, and then infer that (most) unobserved gold bars are yellow.

How does antirealism differ from inductive skepticism? Inductive skeptics observe that T makes true predictions, but do not infer that it will make true predictions. As far as they are concerned, there is no epistemic difference between the belief that unobserved gold bars are yellow and the belief that they are not yellow, and between the prediction that T will make true predictions and the prediction that it will make false predictions. None of the two beliefs and the two predictions is justifiable.

How do participants in the realism debate respond to inductive skepticism or to the problem of induction? Some of them set it aside. They assume that induction is reliable, and then attempt to determine whether realism is tenable. P. D. Magnus and Craig Callender write that "rightly or wrongly, that problem is being bracketed here (assumed 'solved' or 'unresolvable')" (Magnus & Callender, 2004, p. 323). Other participants, such as van Fraassen and Wray, embrace inductive skepticism. Van Fraassen states, "I do not think that there is such a thing as Induction, in any form" (van Fraassen, 2007, pp. 343–344). Wray endorses van Fraassen's position, saying that he accepts "van Fraassen's attack on induction and inductive inference" (Wray, 2021, p. 246).

In my view, embracing inductive skepticism comes with an advantage and disadvantages. The advantage is that they can avoid criticisms from realists. The disadvantages are that they have to give up their positive philosophical views and negative arguments against realism. The disadvantages are not at all appreciated in the literature, so I flesh them out in the next two sections.

6.3 Antirealists' Positive Philosophical Views

In this section, I elucidate the disastrous consequences of inductive skepticism for antirealists' positive philosophical views: constructive empiricism (van Fraassen, 1980, p. 12; Wray, 2012, p. 378), the selectionist explanation (van Fraassen, 1980, pp. 39–40; Wray, 2018, p. 149), the contextual theory (van Fraassen, 1980, Chap. 5), and the permissive view of rationality (van Fraassen, 1989, pp. 171–172; Wray, 2018, pp. 136–140, 2021, p. 245).

Let me begin with constructive empiricism. It holds that "*acceptance of a theory involves as belief only that it is empirically adequate*" and "*Science aims to give us theories which are empirically adequate*" (van Fraassen, 1980, p. 12). Note that constructive empiricism consists of two components: an acceptance component and an aim component. The former concerns what is involved in the acceptance of T, and the latter concerns what science pursues. To accept T is to make "a commitment to confront any future phenomena by means of the conceptual resources of this theory" (van Fraassen, 1980, p. 12). Accordingly, the acceptance component means that when scientists use T for scientific purposes—such as predicting, explaining, and manipulating—they believe that it is empirically adequate, i.e., the acceptance component attributes the belief that T is empirically adequate to scientists, those who use it for scientific purposes (Park, 2022a, p. 60).

What problem does inductive skepticism pose for the acceptance component of constructive empiricism? If constructive empiricists embrace inductive skepticism, they cannot justifiably attribute the belief that T is empirically adequate to scientists, i.e., the acceptance component collapses. Attributing the belief to scientists requires that constructive empiricists randomly select a large number of scientists, investigate whether they hold the belief, and then make the inductive inference about the population of scientists. However, such a method is not available to constructive empiricists who have embraced inductive skepticism.

Let me turn to the aim component that science aims at empirically adequate theories. If inductions are not justifiable, scientists cannot know that they have obtained empirically adequate theories even if they have them in their hands, and thus constructive empiricists owe us an explanation of why they propose that science aims at empirically adequate theories. Do scientists have a method other than induction to know that they have achieved this aim? If so, what is it? In the absence of an answer to this question, constructive empiricists have no basis for their claim that science aims at empirically adequate theories. Scientists will only ask a blunt question to constructive empiricists: "Why do you think science is in the business of pursuing something that scientists cannot know that they have obtained?"

Let me turn to the selectionist explanation of why our theories make true predictions. Wray writes, "Our theories enable us to make accurate predictions, because scientists do not work with theories that do not enable them to make such predictions" (Wray, 2018, p. 149). Wray also states that "van Fraassen's selectionist explanation is superior to the realists' explanation" on the grounds that the former can explain, while the latter cannot, why some successful theories come to be rejected and why two rival theories can be both successful (Wray, 2018, p. 149).

Note that Wray claims that the selectionist explanation can handle more scientific phenomena than can the realist one. Specifically, selectionism can account for, while realism cannot, why some successful scientific theories come to be rejected and why two rival scientific theories can be both successful. In a nutshell, Wray's argument for the selectionist explanation is that since the selectionist explanation has a wider scope than its realist counterpart, we can justifiably believe the selectionist explanation.

How would inductive skeptics respond to Wray's argument for the selectionist explanation? They would point out that it is an inductive argument. More specifically, it is an inference to the best explanation (IBE), and they are more skeptical about IBE than about enumerative induction, given that IBE might involve unobservables while enumerative induction only involves observables (Sect. 1.3.1 of Chap. 1). Consequently, they would reject Wray's assertion that "Our theories enable us to make accurate predictions, because scientists do not work with theories that do not enable them to make such predictions" (Wray, 2018, p. 149). To state it differently, they would reject Wray's selectionist explanation.

Let move on to the contextual theory, according to which "An explanation is an answer to a why-question" (van Fraassen, 1980, p. 134), and "among the scientifically relevant factors, context determines explanatorily relevant ones" (van Fraassen, 1980, p. 126). Van Fraassen puts forward this theory with the express aim of explaining rejections and asymmetries (van Fraassen, 1980, p. 134). To put it briefly, some requests for explanations are rejected in science, and p can explain q, but not vice versa, in science. A philosophical theory of scientific explanation should explain why such phenomena occur in science. Van Fraassen applies the contextual theory to rejections (van Fraassen, 1980, p. 128) and to asymmetries (van Fraassen, 1980, pp. 132–134).

How would inductive skeptics respond to van Fraassen's argument for the contextual theory? They would point out that it is utilizes IBE, a rule of inference about which they are skeptical (Park, 2022b, S313–315). Accordingly, they would reject van Fraassen's assertion that "An explanation is an answer to a why-question" (van Fraassen, 1980, p. 134) and "among the scientifically relevant factors, context determines explanatorily relevant ones" (van Fraassen, 1980, p. 126). To state it otherwise, they would reject the contextual theory.

Let me turn to the two rival conceptions of rationality: the defensive and permissive views. On the defensive view, "what is rational to believe is exactly what one is rationally compelled to believe" (van Fraassen, 1989, p. 171). On the permissive view, "what it is rational to believe includes anything that one is not rationally compelled to disbelieve" (van Fraassen, 1989, pp. 171–172). On the permissive view, "any beliefs (or theories) consistent with the data and the canons of rationality are permissible" (Wray, 2018, p. 245). Van Fraassen (1989, p. 171, 2017, p. 102) and Wray (2018, pp. 136–140, 2022, p. 245) espouse the permissive view over the defensive view.

The permissive view seems to have the following advantage. It allows antirealists to rationally believe that T is empirically adequate without justifying inductions. They can argue that there is no argument that rationally compels them to disbelieve that it is empirically adequate, and thus that they can rationally believe that it is empirically adequate. Those who oppose this attitude toward T would have the heavy burden of presenting an argument that would rationally compel antirealists to disbelieve that T is empirically adequate. Their argument should not be inductive because antirealists do not think that inductive arguments are rationally compelling. It appears that the permissive view is a clever idea to shift the burden of proof to inductive skeptics, those who attempt to dissuade antirealists from believing that T is empirically adequate.

Van Fraassen does not construct any argument for his choice of the permissive view over the defensive one, but Wray does. Wray's argument goes as follows. Suppose that T_1 and T_2 compete with each other. On the defensive view, if it is rational to believe T_1, it is irrational to believe T_2, and vice versa. On the permissive view, by contrast, it is rational for some scientists to believe T_1 and also rational for other scientists to believe T_2. The permissive view "can account for the fact that rational scientists can and do disagree" (Wray, 2021, p. 245). Note that Wray chooses the permissive view over the defensive view on the grounds that the former has a wider explanatory scope than the latter.

How would inductive skeptics respond to Wray's argument for the permissive view? They might accept the premise that the permissive view has a broader explanatory scope than the defensive view, but they would reject the conclusion that "any beliefs (or theories) consistent with the data and the canons of rationality are permissible" (Wray, 2018, p. 245). Their rejection of the conclusion is grounded in the fact that Wray's argument for the permissive view exemplifies IBE.

Moreover, the permissive view is self-defeating. Just as scientific theories, T_1 and T_2, compete with each other, so theories of rationality compete with each other. The permissive and defensive views are rival conceptions of rationality. Just as the permissive view implies that it is rational to believe T_1 but also rational to believe T_2, so it implies that it is rational to accept the permissive view but also rational to accept the defensive view. To state it differently, the permissive view should be permissive not only about rival scientific theories but also about rival theories of rationality. Permissivists cannot accuse defensivists of being irrational for applying the defensive view to the competition between T_1 and T_2 or to the competition between the permissive and defensive views. As a consequence, under the permissive view, it is rational to reject the permissive view and accept the defensive view.

The permissive view also has a pragmatic disadvantage. Suppose that a team of scientists plans to send a rocket to Mars. Carrying out the plan requires that they accurately predict the positions of Mars. All of the team members agree that Mars has been on a certain orbit, o_1, but they disagree with one another on what the future orbit of Mars will be. One member predicts that the future orbit will be o_1 as it used to be, another predicts that it will be o_2, another predicts that it will be o_3, and so on. To make matters worse, they all have adopted the permissive view, and then respectively claim that it is rational to believe that the future orbit will be o_1, o_2, o_3, and so on. Under such circumstances, any teamwork between them would be impossible, and they would be unable to send a rocket to Mars. In short, if scientists embraced the permissive view, science could not exist.

Wray would reply that according to the permissive view, it is rational to believe that the future orbit of Mars will be o_1 as it used to be, but irrational to believe that it will be other than o_1, and thus the permissive view does not make it impossible for scientists to collaborate. Unfortunately, this possible reply creates the following two heavy burdens of proof for Wray.

First, Wray has to show that it is irrational to believe that the future orbit of Mars will be other than o_1. He might construct an argument to show that it is inconsistent with the canons of rationality to believe that the future orbit of Mars will be other than o_1. His argument, whatever it might be, should not be inductive because he has embraced inductive skepticism. It is beyond my imagination how inductive skeptics could show that it is only rational to believe that the future orbit of Mars will be o_1.

Second, Wray has to justify his double standard toward competing theories and competing sets of observational data. On the one hand, he claims that it is rational for some scientists to believe T_1 and also rational for other scientists to believe T_2. On the other hand, he claims that it is rational for some scientists to predict that the future orbit of Mars will be o_1, but irrational for other scientists to predict that it will be other than o_1. It is not clear to me why he applies the permissive view to competing theories, but not to competing sets of observational data.

6.4 Antirealists' Negative Arguments Against Scientific Realism

In this section, I expose the disastrous consequences of inductive skepticism for antirealists' negative arguments against realism: the pessimistic induction, the argument from a bad lot, and the argument from indifference.

Let me begin with the pessimistic induction. It holds that since scientific revolutions have occurred in the past, they will also occur in the future (Wray, 2018, p. 202). The pessimistic induction is vulnerable to David Hume's (1993, pp. 15–25) argument against inductions, which is analyzed in detail in Sect. 1.2 of Chap. 1. Hume would say that the premise that scientific revolutions have occurred in the past is justified, but that the conclusion that scientific revolutions will occur in the future is not. The conclusion is not justified a priori because no contradiction is involved in its negation that scientific revolutions will not occur in the future. Nor is it justified a posteriori because the development of science is not necessarily connected with revolutions. The uniformity principle, which implies that the behavior of science is uniform, is required for the conclusion to be justified. The uniformity principle is not justified a priori because no contradiction is involved in its negation that nature is not uniform. Nor is it justified a posteriori because it is circular to appeal to an induction to justify it. Therefore, the pessimistic induction is not justified.

On the one hand, Wray runs the pessimistic induction against realism. On the other hand, he retreats to inductive skepticism when antirealism is under fire. He is aware that it is incoherent for inductive skeptics to advocate the pessimistic induction. Apropos of this problem, he states that the pessimistic induction raises "a significant challenge for the realist" (Wray, 2021, p. 246). The idea seems to be that he does not believe the pessimistic induction, and that he merely advances it to pose a challenge to realists.

I make the following two critical comments on Wray's response to the foregoing problem. First, Wray (2018, *passim*) writes as if he believes the pessimistic induction. For example, he says that "theories are apt to continue to be discarded indefinitely into the future" (Wray, 2018, p. 202). As I (Park, 2017, p. 383, 2022a, p. 51) have already pointed out, Moore's paradox might arise when antirealists state that p despite the fact that they do not believe that p. Second, realists might play a similar game. When the no-miracles argument is under fire, they can say that they advanced it not because they believed it but because they wanted to present a challenge to antirealists. In consequence, both realists and antirealists can be interpreted as advancing arguments to each other without believing them.

This conclusion has the following problematic implications. Realists would take the pessimistic induction less seriously on the grounds that those who put it forward do not believe it. Antirealists would take the no-miracles argument less seriously on the grounds that those who put it forward do not believe it. In addition, observers of the realism debate, who believe that both realists and antirealists advance arguments to each other without believing them, would take the realism debate less seriously.

Let me turn to the argument from a bad lot. Users of IBE contend that T, which best explains relevant facts, is true. According to the argument from a bad lot, the belief of T requires the privilege principle that T is more likely to be chosen from a good lot than from a bad lot. This argument can be summarized as follows:

T might be the best of a bad lot.
∴ The privilege principle is required for the belief of T.

In other words, since T might be false, users of IBE should justify the privilege principle before they believe T. Wray claims that the argument from a bad lot is "far more threatening to scientific realism" than is the pessimistic induction (Wray, 2021, p. 246).

How would inductive skeptics criticize the argument from a bad lot? They might be interested in whether it is deductive or inductive. On the one hand, if it were deductive, they would reject its premise on the grounds that it assumes the point that needs to be justified. On the other hand, if it were inductive, they would reject it on the grounds that inductions are in general not justifiable. Either way, they would reject it.

This dilemma entitles inductive skeptics to reject the argument from a bad lot without determining whether it is deductive or inductive. However, it is an intrinsically interesting issue whether it is deductive or inductive. Determining whether it is deductive or inductive requires that we closely investigate the nature of the premise and the conclusion. Are the premise and the conclusion possibility, descriptive, or normative statements?

The premise is a possibility statement, i.e., it is a possibility statement that T might be the best of a bad lot. Inductive skeptics would accept the possibility statement, given that in general, they accept possibility statements. For example, they accept the possibility statement that the conclusions of inductions might be false (Sect. 1.2 of Chap. 1). It would involve a double standard for them to accept the possibility statement that the conclusions of inductions might be false but to reject the possibility statement that T might be the best of a bad lot.

What is the nature of the conclusion that the privilege principle is required for the belief of T? It is a normative statement, asserting that believers of T ought to justify the privilege principle. Cognitivism and noncognitivism in metaethics hold that a normative statement expresses a cognitive state and a noncognitive state, respectively. Thus, cognitivism affirms, while noncognitivism denies, that the conclusion has a truth-value. If cognitivism is true, the conclusion has a truth-value, and the argument from a bad lot is either deductive or inductive. By contrast, if noncognitivism is true, the conclusion does not have a truth-value, and the argument from a bad lot is not an argument in the first place, although it is called "the argument from a bad lot," and thus it is neither deductive nor inductive.

How would inductive skeptics respond to the argument from a bad lot under the framework of cognitivism? They would reject it on the grounds that it is inductive. It is inductive, given that the premise that T might be the best of a bad lot is compatible with the negation of the conclusion that the privilege principle is not required for the belief of T, i.e., given that the premise does not entail the conclusion.

How would inductive skeptics respond to the conclusion of the argument from a bad lot under the framework of noncognitivism? They might have a positive emotion toward the privilege principle, and say that it is required for the belief of T to be justified. Or they might have a negative emotion toward the privilege principle and say that it is not required for the belief of T to be justified. These two rival options are available to inductive skeptics.

What does the existence of these rival options show? Under the framework of noncognitivism, van Fraassen and Wray have the burden of explaining why they have a positive emotion toward the privilege principle and say that the privilege principle is required for the belief of T to be justified. After all, their fellow inductive skeptics, who also operate under noncognitivism, might have a negative emotion toward the privilege principle, and say that the privilege principle is not required for the belief of T to be justified.

Let me move on to the argument from indifference. There are two versions of it in the literature. Van Fraassen's version holds that "there are many theories, perhaps never yet formulated but in accordance with all evidence so far, which explain at least as well as the best we have now" (van Fraassen, 1989, p. 146). Therefore, the best formulated theory is unlikely to be true, i.e., "it must seem very improbable to me that it is true" (van Fraassen, 1989, p. 146). In short, the premise is that there *are* many unformulated theories, and the conclusion is that the best formulated theory is unlikely to be true.

How would inductive skeptics respond to this version of the argument from indifference? They would point out that the premise is not justified. Van Fraassen has not observed any unformulated theory, but claims that there are many unformulated theories. To use an analogy, he has not seen any blue gold bar, but claims that there are many blue gold bars. Such inferences would not sound convincing to anyone, to say nothing of inductive skeptics.

Van Fraassen would reply that he observed the instances in which scientists formulated new theories that were previously unformulated, and thus that he is entitled to infer that there are many unformulated theories. However, this argument is inductive, and thus it would not sound convincing to inductive skeptics. It is not clear to me how van Fraassen can justify the premise that they are many unformulated theories without relying on an induction.

James Ladyman et al.'s variant of the argument from indifference holds that "the possibility that there may be equally good rival theories to T already suffices to make an ampliative step from the evidence to T unwarranted" (Ladyman et al., 1997, p. 309). Note that the premise is that it is *possible* that there might be many unformulated theories, and that the conclusion is that T is *unwarranted.*

How would inductive skeptics respond to this version of the argument from indifference? The conclusion is a normative statement, asserting that believers of T ought not to believe T. Under the framework of cognitivism, the conclusion expresses a belief, and the argument is inductive, given that the premise is compatible with the negation of the conclusion, i.e., no contradiction is involved in the assertion that there might be many unformulated theories, but that it is warranted. As a result, inductive skeptics would reject Ladyman et al.'s version of the argument from indifference.

Under the framework of noncognitivism, by contrast, the conclusion expresses, for example, an emotion. The alleged argument is not an argument in the first place, and hence it would neither be deductive nor inductive. Some noncognitivists might have a negative emotion toward T, and say that it is unwarranted. By contrast, other noncognitivists might have a positive emotion toward T, and say that it is warranted. It follows that under the framework of noncognitivism, Ladyman et al. have the burden of explaining why they have a negative emotion toward T and say that it is unwarranted.

To sum up, it is costly for antirealists to retreat to inductive skepticism. In my view, they would be better off if they resisted inductive skepticism. But how can they resist inductive skepticism and realism at the same time? My answer to this question can be found in the next section.

6.5 Does the Local Theory Help Scientific Realism?

In this section, I defend negative answers to the following questions. Does the local theory undercut antirealism and support realism? If antirealists embrace the local theory, should they reject antirealism and accept realism?

The local theory holds that a local uniformity principle licenses a set of inductions in a domain. For example, the local uniformity principle that planets move around the sun licenses the induction that since Venus moved around the sun, it will move around the sun, and the induction that since Mercury moved around the sun, it will move around the sun. Different local uniformity principles license different sets of inductions in different domains. The local theory contrasts with the global theory, according to which the global uniformity principle licenses all (cogent) inductions in all domains (Sect. 3.2 of Chap. 3).

Shech thinks that the local theory undermines antirealism, stating that "whether or not a particular inductive argument works, whether inductive inferences are warranted, does not depend on whether said claims are about observables or unobservables" (Shech, 2021, p. 229). His idea is that an inductive inference is justified, whether it is from observables to observables or to unobservables, insofar as a local uniformity principle licenses it. Therefore, it is problematic for antirealists to accept inductive inferences from observables to observables but to reject inductive inferences from observables to unobservables.

In response, Wray (2021, p. 246) retreats to inductive skepticism, approvingly citing van Fraassen's statement: "I do not think that there is such a thing as Induction, in any form" (van Fraassen 2007, pp. 343–344). In my view, it is understandable that antirealists retreat to inductive skepticism when antirealism is under fire. After all, once they retreat to inductive skepticism, they can avoid the attacks on antirealism. However, this single advantage of retreating to inductive skepticism comes with multiple disadvantages. As noted in Sects. 6.3 and 6.4 above, antirealists would have to give up their positive philosophical views and their negative arguments against realism.

If I were an antirealist, I would not retreat to inductive skepticism; I would rather criticize the local theory that Shech relies on to criticize antirealism. In Sect. 3.3 of Chap. 3, I raised several objections to the local theory. The most notable one is that it is to confuse a means and an end to say that a local uniformity principle is required for a set of inductions in a domain to be justified. The inevitable consequence of confounding the means and the end is skepticism about both a set of inductions and a local uniformity principle. In short, the local theory creates rather than solves the problem of induction.

This observation on the local theory has two unsavory implications for Shech. First, he cannot appeal to the local theory to argue against antirealism according to which inductive inferences from observables to observables are legitimate while inductive inferences from observables to unobservables are illegitimate. After all, the local theory implies that inductive inferences, whether they are from observables to observables or to unobservables, are not justifiable. Second, Shech advocates local realism, a variant of realism (Shech, 2021, p. 235). Any variant of realism is a product of inductions. However, the local theory implies that inductions are not justifiable, and thus that local realism is not justifiable. In my view, Shech cannot have both the local theory and local realism.

Suppose for the sake of argument that the local theory does not lead to inductive skepticism, and that it rather solves the problem of induction, i.e., that it shows that inductions are justified, as Norton and Shech contend. Further suppose that antirealists embrace the local theory. An interesting question arises. Should antirealists give up antirealism and accept realism? My answer is "No." Just because they embrace the local theory, it does not follow that they have to give up antirealism and embrace realism. They might take, by fiat, the arbitrary position that we can know that local uniformity principles license the inductive inferences from observables to observables, but that we cannot know that local uniformity principles license the inductive inferences from observables to unobservables. In my view, taking this arbitrary position is better than retreating to inductive skepticism. After all, retreating to inductive skepticism comes with the costs mentioned in Sects. 6.3 and 6.4 above.

Can antirealists justify the arbitrary position? They can argue that observational beliefs were stable while theoretical beliefs were unstable in the history of science, and thus that we can know that local uniformity principles license the inductive inferences from observables to observables while we cannot know that local uniformity principles license the inductive inferences from observables to unobservables. Admittedly, this argument needs to be bolstered to transform the arbitrary position to a justified one. I leave the task of bolstering it to future antirealists.

6.6 Conclusion

If antirealists embrace inductive skepticism, they can avoid the criticisms against antirealism. However, this sole advantage comes with the disadvantages of losing their positive philosophical views and negative arguments against realism. Antirealists would be better off if they rejected the local theory on the grounds that it is vulnerable to my several objections to it in Sect. 3.3 of Chap. 3. Even if the local theory solves the problem of induction, antirealists are better off if they take the arbitrary position than if they retreat to inductive skepticism. This chapter has the following message to antirealists: "Inductive skepticism incurs cognitive famine."

References

Hume, D. (1993). *An enquiry concerning human understanding*. Hackett Publishing.

Ladyman, J., Douven, I., Horsten, L., & van Fraassen, B. (1997). A defense of van Fraassen's critique of abductive inference: Reply to Psillos. *The Philosophical Quarterly, 47*(188), 305–321.

Magnus, P. D., & Callender, C. (2004). Realist ennui and the base rate fallacy. *Philosophy of Science, 71*(3), 320–338.
Norton, J. (2003). A material theory of induction. *Philosophy of Science, 70*(4), 647–670.
Park, S. (2017). Understanding without justification and belief? Principia. *An International Journal of Epistemology, 21*(3), 379–389.
Park, S. (2022a). *Embracing scientific realism*. Springer.
Park, S. (2022b). The contextual theory of explanation and inference to the best explanation. *Axiomathes, 32*(S2), 311–326.
Shech, E. (2021). Darwinian-selectionist explanation, radical theory change, and the observable-unobservable dichotomy. *International Studies in the Philosophy of Science, 34*(4), 221–241.
van Fraassen, B. (1980). *The scientific image*. Oxford University Press.
van Fraassen, B. (1989). *Laws and symmetry*. Oxford University Press.
van Fraassen, B. C. (2007). From a view of science to a new empiricism. In B. Monton (Ed.), *Images of empiricism: Essays on science and stances, with a reply from Bas C. van Fraassen* (pp. 337–383). Oxford University Press.
van Fraassen, B. (2017). Misdirection and misconception in the scientific realism debates. In E. Agazzi (Ed.), *Varieties of scientific realism: Objectivity and truth in science* (pp. 95–108). Springer.
Wray, B. (2012). Epistemic privilege and the success of science. *Noûs, 46*(3), 375–385.
Wray, K. (2018). *Resisting scientific realism*. Cambridge University Press.
Wray, K. (2021). Induction, rationality, and the realism/antirealism debate: A reply to Shech. *International Studies in the Philosophy of Science, 34*(4), 243–247.

Chapter 7
The Pessimistic Inductions vs. the Counter-Inductions

Abstract This chapter defends counter-inductions against three pessimistic inductions: the classic pessimistic induction (CPI), the problem of unconceived alternatives (PUA), and the problem of misleading evidence (PME). The counter-induction against CPI holds that since earlier theories were false, it is likely that contemporary theories are true. The counter-induction against PUA holds that since past scientists could not exhaust sets of alternatives, it is likely that current scientists can exhaust sets of alternatives. The counter-induction against PME holds that since the evidence for past theories was misleading, it is likely that the evidence for current theories is truth-leading.

Keywords Classic pessimistic induction · Problem of misleading evidence · Problem of unconceived alternatives

7.1 Introduction

Pessimists have developed diverse pessimistic inductions (Park, 2022, pp. 28–40). This chapter focuses on three of them: the classic pessimistic induction (CPI), the problem of unconceived alternatives (PUA), and the problem of misleading evidence (PME). CPI holds that the falsity of past theories indicates the falsity of current theories (Laudan, 1977, p. 126). PUA holds that past scientists' inability to exhaust sets of alternatives indicates current scientists' inability to exhaust sets of alternatives (Stanford, 2006, p. 20). PME holds that the misleadingness of the evidence for past theories indicates the misleadingness of the evidence for current theories (Frost-Arnold, 2019, p. 912).

This chapter aims to defend counter-inductions against the three pessimistic inductions. In Sect. 7.2, I distinguish between induction and counter-induction, and then argue that induction is rational in some cases but counter-induction is rational in other cases. In Sect. 7.3, I defend the counter-induction against CPI that the

S. Park, *Induction, Science, and Morality*, Synthese Library 520,
https://doi.org/10.1007/978-3-032-00032-3_7

falsity of past theories indicates the (approximate)[1] truth of current theories (Park, 2018a, p. 335). In Sect. 7.4, I defend the counter-induction against PUA that past scientists' inability to exhaust sets of alternatives indicates current scientists' ability to exhaust sets of alternatives. In Sect. 7.5, I defend the counter-induction against PME that the misleadingness of the evidence for past theories indicates the truth-leadingness of the evidence for current theories. In Sect. 7.6, I reply to a possible move that pessimists might make regarding my two arguments for preferring the three counter-inductions to the three pessimistic inductions.

The three counter-inductions, if correct, rebut the popular view of the history of science, viz., it is the graveyard of dead theories. As Moti Mizrahi (2016, pp. 263–264) observes, the word "graveyard" or "cemetery" is used to characterize the history of science by many philosophers, such as Isabelle Stengers (2000, p. 31), Peter Lipton (2005, p. 1265), Anjan Chakravartty (2008, p. 152), Frost-Arnold (2011, p. 1138), Hasok Chang (2011, p. 426), Roy Sorensen (2013, p. 31), Daniel Graham (2013, p. 22), David Harker (2015, p. 92), and Peter Harrison (2015, p. 181). In the end, readers will see that it is more reasonable to view the history of science as the birthplace of true theories than as the graveyard of dead theories.

7.2 Induction vs. Counter-Induction

In this section, I make a distinction between induction and counter-induction, argue that induction is rational in some cases while counter-induction is rational in other cases, and preempt a possible attempt to resolve a stalemate between pessimists and counter-inductivists.

We are inductivists in some cases but counter-inductivists in other cases. For example, we make an inductive inference that because water has flowed downward in the past, it will flow downward in the future. We make a counter-inductive inference, for example, that because Alice has made mistakes in the past, she will not make a similar mistake in the future. None of us is always an inductivist, and none of us is always a counter-inductivist.

Scientists often experience failures before they make an achievement. Pessimists and counter-inductivists have opposing views about scientists' previous failures. For pessimists, previous failures indicate forthcoming failures. The more scientists have failed, the more probable it is that they will fail next time. It follows that the "phrase 'the achievement after trials and errors' is not in pessimists' dictionary" (Park, 2017, p. 217). For counter-inductivists, by contrast, previous failures indicate forthcoming success. The more scientists have failed, the more probable it is that they will succeed next time. It follows that the above phrase is in counter-inductivists' dictionary.

Let me apply these two opposing views about scientists' failures to the following example of a funding agency. Suppose that you are the director of a funding agency,

[1] I drop "approximate" henceforth.

that you have funded a research team, and that the research team has failed many times to achieve its research goal. Should you keep funding the research team? If you were a pessimist, you would predict that the research team will continue to fail, and you would thus stop funding them. In contrast, if you were a counter-inductivist, you would predict that the research team will succeed, and you would thus keep funding them (Park, 2017, p. 222).

No counter-inductivist would make the strong claim that all counter-inductions are rational, just as no inductivist would make the strong claim that all inductions are rational. Counter-inductivists would rather make the weak claim that some counter-inductions are rational while other counter-inductions are irrational, just as inductivists would rather make the weak claim that some inductions are rational while other inductions are irrational. Consequently, to present some irrational counter-inductions is not to refute counter-inductivists' position. For example, alchemists failed many times to make gold out of lead, and it would have been irrational for alchemists to think that since they failed many times, they would succeed next time. The existence of such counter-inductions does not refute counter-inductivists' position (Park, 2018a, p. 332).

Pessimists and counter-inductivists would challenge each other to justify their inferences. Pessimists would ask counter-inductivists the following question. How can past failures be evidence for a prediction of future success? By contrast, counter-inductivists would ask pessimists to the following question. How can past failures be evidence for a prediction of future failure? Consequently, if counter-inductivists have the burden of justifying counter-induction, pessimists also have the burden of justifying induction. It is unfair to claim that counter-inductivists have the burden of justifying counter-induction, while inductivists do not have the burden of justifying induction (Park, 2018a, pp. 332–334).

How can we resolve the stalemate between pessimists and counter-inductivists? Consider the set of all episodes from the history of science. In some of those episodes, scientists continued to fail without reaching success, while in other episodes, they succeeded after they had failed. We might randomly select a large number of sample episodes from the population of all historical episodes, and then suggest that induction is reliable on the grounds that the set of sample episodes in which scientists continued to fail is larger than the set of sample episodes in which scientists succeeded after they had failed, or suggest that counter-induction is reliable on the grounds that the former set is smaller than the latter set.

Unfortunately, the foregoing random sampling method does not work. Suppose that the set of episodes in which scientists continued to fail was larger than the set of episodes in which scientists succeeded after they had failed. Counter-inductivists would argue that since the former set was larger than the latter set, the set of episodes in which scientists will continue to fail will be smaller than the set of episodes in which scientists will succeed after they fail. In other words, counter-inductivists would argue that since scientists were more likely to continue to fail after they had failed, they will be more likely to succeed after they fail in the future. Therefore, the history of science cannot be an arbiter to break the stalemate between pessimists and counter-inductivists.

7.3 The Counter-Induction Against CPI

In this section, I introduce a counter-induction against CPI, and then present two arguments for preferring the counter-induction to CPI. I also explore the possible dialectics between pessimists of CPI and counter-inductivists.

CPI holds that most "of the past theories of science are already suspected of being false; there is presumably every reason to anticipate that current theories of science will suffer a similar fate" (Laudan, 1977, p. 126). On this account, the fact that the caloric theory was false is evidence that the kinetic theory is false. CPI is built upon what I earlier called "proportional pessimism" (Park, 2016, p. 836), according to which the probability that a theory is false is proportional to the quantity of its false forerunners, i.e., the more false forerunners it has, the more probable it is that it is false.

Proportional pessimism should be obvious. Suppose that T_1, T_2, and T_3 are successive rivals. T_1 and T_2 are obsolete theories, while T_3 is a current theory. According to CPI, the falsity of T_1 was evidence that T_2 was false, and the falsity of T_2 was evidence that T_3 is false. It follows that T_1 was most probable, and T_3 is least probable. For example, Aristotle's mechanics was most probable, and Einstein's mechanics is least probable. Keep this implication of proportional pessimism in mind, as it will be utilized below.

Let me unpack the counter-induction against CPI. It holds that because earlier theories were false, it is likely that contemporary theories are true (Park, 2018a, p. 335). For example, the fact that the caloric theory was false is evidence that the kinetic theory is true. The more false forerunners a contemporary theory has, the more likely it is that it is true. The premise of the counter-induction is the same as the premise of CPI, but its conclusion is the opposite of the conclusion of CPI. Pessimists of CPI and counter-inductivists draw opposite conclusions from the same premise.

What if a current theory, say, the kinetic theory, is discarded? Pessimists might excitedly say that they were right, while counter-inductivists were wrong, about the kinetic theory, i.e., that the kinetic theory was false as they thought and contrary to what counter-inductivists thought. However, counter-inductivists would retort that strictly speaking, the demise of the kinetic theory is compatible with the counter-induction, pointing out that it does not claim in the first place that the kinetic theory was definitely true, but rather that it was probably true. Moreover, counter-inductivists would say that the demise of the kinetic theory indicates more strongly that its successor is true than the demise of the caloric theory indicated that the kinetic theory was true. Therefore, scientific revolutions will only intensify counter-inductivists' conviction that new theories are true (Park, 2018a, p. 335).

Pessimists and counter-inductivists would challenge each other to justify their inferences. Pessimists would ask counter-inductivists the following questions. What justifies the counter-inductive inference from the falsity of the caloric theory to the truth of the kinetic theory? How can the fact that the caloric theory was false be an indication that the kinetic theory is true? However, counter-inductivists would ask pessimists the following corresponding questions. What justifies the inductive inference from the falsity of the caloric theory to the falsity of the kinetic theory? How can the fact that the caloric theory was false be an indication that the kinetic theory is also false? In short, if counter-inductivists have the burden of justifying the counter-inductive inference, pessimists also have the burden of justifying the inductive inference.

I offer the following two arguments for preferring the counter-induction to CPI. The first one is that as some philosophers (Musgrave, 1985, p. 211; Leplin, 1997, p. 141) point out, the evidence for present theories is more powerful than that for past theories. To put it otherwise, current scientists have more empirical evidence for their theories than past scientists had for their theories. Realists and antirealists agree on whether present theories are more probable than past ones, although they disagree on whether present theories are successful enough to be true, as Florian Müller (2015, p. 406) and Mario Alai (2017, p. 3283) observe. Consequently, counter-inductivists do not commit the gambler's fallacy when they maintain that unlike past theories, current ones are true.

However, this argument for preferring the counter-induction to CPI would be rejected by K. Brad Wray (2013, p. 4327) and Mario Alai (2017, p. 3282). They run what I earlier called "the relative pessimistic induction," according to which T_2 was ousted although it was more successful than T_1, so T_3 will be ousted although it is more successful than T_2 (Park, 2019a, p. 98). The relative pessimistic induction is so named because the premise and the conclusion are both relative theses resulting from the comparison of the success of successive theories.

The relative pessimistic induction is fallacious (Park, 2019a, pp. 98–100). To use an analogy, suppose that t_1 and t_2 are past points in time, that t_3 is the present point in time, and that Diana was a child at t_1 and t_2. It is a fallacious inference that since she was a child even if time passed from t_1 to t_2, it is likely that she would also be a child even if time passed from t_2 and t_3. After all, she might be an adult at t_3. A similar objection can be raised against the relative pessimistic induction. It is a fallacious inference that since scientists did not obtain a true theory when they replaced T_1 with T_2, it is likely that they did not obtain a true theory either when they replaced T_2 with T_3. After all, T_3 might be true unlike T_1 with T_2.

Different theories are successful to different degrees. Those theories whose degrees of success are below a certain point—called "a flipping point" (Park, 2017, p. 217)—are likely to be false, but those theories whose degrees of success are above the flipping point are likely to be true. In general, a flipping point is the point that separates the unobserved objects that are similar to observed objects from the unobserved objects that are dissimilar to observed objects. An unobserved object which is below the point is similar to observed objects, while an unobserved object which is above the point is dissimilar to observed objects. Counter-inductivists believe, for example, that the degrees of success of the humoral and miasma theories were below the flipping point, so they were likely to be false, but that the degree of the success of the germ theory is above the flipping point, so it is likely to be true.

The relative pessimistic induction does not take into account the existence of the flipping point, and thus it is fallacious.

To go further, counter-inductivists would run a counter-induction against the relative pessimistic induction. The counter-induction holds that since scientists did not go above the flipping point when they moved from T_1 to T_2, it was likely that they have gone above the flipping point when they moved from T_2 to T_3. For example, since scientists did not obtain a true theory when they moved from the humoral theory to the miasma theory, it is likely that they have obtained a true theory when they moved from the miasma theory to the germ theory (Park, 2018a, p. 339).

The second argument for preferring the counter-induction to CPI is that CPI is incoherent. What would be the reason for thinking that CPI is incoherent? Compare the probabilities of T_1, T_2, and T_3. Proportional pessimism implies, as we noted earlier in this section, that T_1 is most probable and T_3 is least probable. This implication of proportional pessimism contradicts the corollaries of both assumptions that the number of alternatives is finite and infinite, as explicated below.

Suppose first that a set of alternatives is finite. As more and more alternatives are eliminated from the set of alternatives, the number of remaining alternatives decreases. The smaller the number of remaining alternatives is, the more likely it is that they are true (Ruhmkorff, 2013, p. 412; Park, 2016, p. 840). Accordingly, T_1 is least probable and T_3 is most probable. This corollary contradicts the implication of proportional pessimism that T_1 is most probable and T_3 is least probable (Park, 2016, p. 839).

Suppose now that a set of alternatives is infinite. No matter how finitely many alternatives were eliminated from the set of alternatives, there would still be infinitely many members in the set of alternatives, and they would all be 0% probable. This corollary contradicts the implication of proportional pessimism that T_1 is most probable and T_3 is least probable (Park, 2016, p. 839).

How would pessimists respond to the foregoing two arguments for preferring the counter-induction to CPI? They might go on the offensive, pointing out that the counter-induction requires the assumption that the number of alternatives is finite. If the number of alternatives is finite, current theories are more probable than their predecessors. This corollary coheres with the counter-induction. However, if the number of alternatives is infinite, past and current theories are all 0% probable. This corollary is incompatible with the counter-induction. Consequently, counter-inductivists need to justify the assumption that the number of alternatives is finite.

My response to this request is as follows. As Samuel Ruhmkorff (2011) observes, just because existing theories were displaced by new ones in the history of science, it does not follow that the number of alternatives is infinite. In other words, the history of science does not tell us anything about whether the number of theories is finite or infinite. Relatedly, it is a hasty generalization to say that since we have observed some alternatives in the history of science, there are infinitely many alternatives (Park, 2018b, p. 14). It appears, therefore, that the only conclusion that we can draw is that the number of alternatives is finite.

What if pessimists disagree with this conclusion? I invite them to consider the following episode of cloning a sheep. Ian Wilmut's research team failed many times in the 1990s before they finally succeeded in cloning a sheep (Park, 2017, p. 216). The number of *unperformed experimental failures* to clone a sheep was either finite or infinite. If it had been finite, it would have been possible for the research team to succeed in cloning a sheep. If it had been infinite, it would have been impossible for the research team to succeed in cloning a sheep. After all, no matter how finitely many times they failed to clone a sheep, there would still have been infinitely many members in the set of infinitely many unperformed experimental failures. The research team actually succeeded in cloning a sheep, thereby proving that the number of unperformed experimental failures was finite.

Let me move on to another episode. In the mid-nineteenth century, Semmelweis set out to decrease the high mortality rate of delivering women in a hospital in Vienna (Hempel, 1966, pp. 3–6). Before he obtained a successful hypothesis, he tested five hypotheses in vain, i.e., he eliminated five unsuccessful hypotheses from a set of unsuccessful hypotheses. The number of *unsuccessful hypotheses* was either finite or infinite. If it had been finite, it would have been possible for Semmelweis to obtain the successful hypothesis. If it had been infinite, it would have been impossible for him to obtain the successful hypothesis. After all, no matter how many finitely many unsuccessful hypotheses were removed, there would still have been infinitely many members in the set of infinitely many unsuccessful hypotheses. Semmelweis actually obtained the successful hypothesis, thereby proving that the number of unsuccessful hypotheses was finite.

Imagine that pessimists were with Semmelweis right before he tested the successful hypothesis. Pessimists predicted that it would be unsuccessful on the grounds that he had tested five unsuccessful hypotheses and the number of unsuccessful hypotheses was infinite. We can see in retrospect that pessimists were wrong (Park, 2018a, pp. 330–331). A similar story can be given regarding the success of Wilmut's research team above. Such historical episodes would entitle us to make the following list of pessimists' mistakes:

The List of Pessimists' Mistakes

- Pessimists would have been wrong about Wilmut's research project.
- Pessimists would have been about Semmelweis's research project.

This list can be expanded *ad nauseam* with other historical episodes in which scientists created new objects after failing many times and obtained successful hypotheses after failing many times. Using this list, we can construct the pessimistic induction against pessimists that since pessimists would have been wrong about scientists' old research projects of creating new objects and obtaining successful hypotheses, it is likely that they are also wrong about scientists' new research projects of creating new objects and obtaining successful hypotheses.

Pessimists cannot appeal to the list of their mistakes above to run a counter-induction for themselves. Specifically, they cannot say that since they would have been wrong about scientists' old research projects, it is likely that they are right about scientists' new research projects. After all, this counter-induction requires the assumption that the number of the ways they can be wrong about scientists' research projects is finite. If it were finite, it would be possible for them to be right about scientists' new research projects. If it were infinite, it would be impossible for them to be right about scientists' new research projects. After all, no matter how many finite times pessimists were wrong about scientists' old research projects, there would still be infinitely many members in the set of alternative ways they can be wrong about scientists' research projects, and thus it would be 0% probable that they will be right about scientists' new research projects.

If pessimists believe that the number of the ways they can be wrong about scientists' research projects is finite, they should also believe that the number of the ways scientists can be wrong about the world is finite, i.e., that the number of alternative scientific theories is finite. It is a double standard for them to say that the number of the ways they can be wrong about scientists' research projects is finite and thus they have a chance to be right about scientists' research projects, while the number of the ways scientists can be wrong about the world is infinite and thus scientists do not have a chance to be right about the world.

What if pessimists accept that the number of alternative scientific theories is finite? It would not be a double standard for them to say that the number of the ways they can be wrong about scientists' research projects is finite. Hence, they could run a counter-induction for themselves, i.e., they could say that since they would have been wrong about scientists' old research projects, it is likely that they are right about scientists' new research projects. If they run the counter-induction for themselves, however, they should also run the counter-induction against CPI. After all, it would be a double standard for them to embrace the counter-induction for themselves but resist the counter-induction against CPI.

Let me construct a different kind of argument, a modus tollens, for the assumption that the number of alternatives is finite. Consider that the caloric theory of heat ran into serious anomalies, and that scientists attempted to develop an alternative to it in the nineteenth century. How probable was it that they would select the kinetic theory of heat from the set of alternatives of heat? If the number of alternatives of heat had been finite, it would not have been 0% probable. If the number of alternatives of heat had been infinite, it would have been 0% probable. Scientists actually selected the kinetic theory from the set of alternatives of heat, thereby proving that it was not 0% probable, and that the number of alternatives of heat was finite.

Let me now construct a reduction ad absurdum for the assumption that the number of alternatives is infinite. As Carlo Alvaro (2021, p. 5) observes, a contradiction can be derived from the idea that the number of past years is infinite. His observation can be applied to the present issue of whether the number of alternatives is infinite. Consider a set of infinitely many alternatives. Even if half of the members were removed from the set, there would still be infinitely many alternatives in the set. Therefore, infinity minus infinity equals infinity. Now, if all the members were removed from the set, there would be no alternative in the set. Therefore, infinity minus infinity equals zero. But how can infinity minus infinity be both infinity and zero? Since the contradiction can be derived from the assumption that the number of alternatives is infinite, the number of alternatives is finite.

7.4 The Counter-Induction Against PUA

In this section, I construct a counter-induction against PUA and sketch the possible dialectics between pessimists of PUA and counter-inductivists.

PUA (Stanford, 2006, p. 20) holds that because past scientists could not exhaust sets of alternatives, it is likely that current scientists cannot exhaust sets of alternatives either. What would be the reason for thinking that like past scientists, current scientists cannot exhaust sets of alternatives? Kyle Stanford's answer is that the space of alternatives "appears to be indeterminate and unbounded" (Stanford, 2006, p. 133). To put it differently, the number of alternatives is infinite. It follows that like past scientists, current scientists are "creatures whose cognitive constitutions are not well suited to the task of exhausting the kinds of spaces of serious candidate theoretical explanations from which our scientific theories are drawn" (Stanford, 2006, p. 45). Note that PUA does not make the agnostic claim that we do not know whether current scientists can or cannot exhaust sets of alternatives, but rather the pessimistic claim that current scientists cannot. It follows that pessimists advocating PUA have the burden of establishing that PUA is correct while the counter-induction against it is incorrect.

Proportional pessimism is also built into PUA. According to PUA, the failure of scientists of T_1 to conceive of T_2 indicated that scientists of T_2 could not conceive of T_3, and the failure of scientists of T_2 to conceive of T_3 indicated that scientists of T_3 cannot conceive of T_4. Accordingly, the failure of scientists of T_1 to conceive of T_2 and the failure of scientists of T_2 to conceive of T_3 jointly indicated more strongly that scientists of T_3 cannot conceive of T_4 than the failure of scientists of T_1 to conceive of T_2 indicated that scientists of T_2 could not conceive of T_3. In a nutshell, scientists of T_1 were most likely to exhaust a set of alternatives, and scientists of T_3 are least likely to exhaust the set of alternatives. Keep this implication of proportional pessimism in mind, as it will be invoked below.

The counter-induction against PUA holds that because past scientists could not exhaust sets of alternatives, it is likely that current scientists can exhaust sets of alternatives. The premise of the counter-induction is the same as the premise of PUA, but its conclusion is the opposite of the conclusion of PUA. Pessimists of PUA and counter-inductivists draw opposite conclusions from the same premise.

What if a current theory, say, the kinetic theory, is displaced by an alternative? Pessimists of PUA might excitedly say that current scientists could not exhaust a set of alternatives to the kinetic theory, as they thought and contrary to what counter-inductivists thought, i.e., that they were right and counter-inductivists were wrong about current scientists' cognitive ability to exhaust the set of alternatives. However, counter-inductivists would retort that strictly speaking, this failure is compatible with the counter-induction, pointing out that it does not claim in the first place that current scientists *definitely* had the cognitive ability, but rather that they were *likely* to have the cognitive ability. Counter-inductivists would add that the fact that current scientists did not have the cognitive ability makes it more likely that future scientists will have the cognitive ability. Therefore, pessimists of PUA and counter-inductivists would remain in a stalemate.

How can we resolve the stalemate in favor of counter-inductivists? In Sect. 7.3 above, I presented two arguments to resolve the stalemate between pessimists of CPI and counter-inductivists. Those arguments can also be used to resolve the stalemate between pessimists of PUA and counter-inductivists. Let me spell out one by one how they can be used.

The first argument is that the evidence for new theories is more extensive than that for old theories. The more observational knowledge scientists have, the more likely it is that they can exhaust sets of alternatives, because the additional knowledge can be used to eliminate false alternatives. For example, Galileo observed the moons of Jupiter, while Ptolemy did not. Galileo had more observational knowledge than Ptolemy. With the additional observational knowledge, Galileo could eliminate the Ptolemaic theory from the set of alternatives, and as a result the number of remaining alternatives decreased. The fewer members there are in a set of alternatives, the more likely it is that scientists can exhaust the set of alternatives. Accordingly, counter-inductivists do not commit the gambler's fallacy when they maintain that unlike past scientists, current scientists can exhaust sets of alternatives.

The second argument is that PUA is incoherent. Compare the cognitive capacity of scientists of T_1, scientists of T_2, and scientists of T_3 to exhaust a set of alternatives. Proportional pessimism implies, as we noted earlier in this section, that scientists of T_1 are most likely to exhaust the set of alternatives, and that scientists of T_3 are least likely to exhaust the set of alternatives. This implication of proportional pessimism contradicts the corollaries of both assumptions that the number of alternatives is finite and infinite, as shown below.

Suppose first that the set of alternatives is finite. As more and more alternatives are eliminated from the set of alternatives, the number of remaining alternatives decreases. The smaller the number of remaining alternatives, the more likely it is that scientists can exhaust the set of alternatives. Accordingly, scientists of T_1 are least likely to exhaust a set of alternatives, and scientists of T_3 are most likely to exhaust the set of alternatives. This corollary contradicts the implication of proportional pessimism that scientists of T_1 are most likely to exhaust the set of alternatives, and that scientists of T_3 are least likely to exhaust the set of alternatives.

Suppose now that the set of alternatives is infinite. No matter how finitely many alternatives were removed from the set of alternatives, there would still be infinitely many members in the set of alternatives. Therefore, it would be equally unlikely that scientists of T_1, T_2, and T_3 can exhaust the set of alternatives, i.e., none of them would have a better chance to exhaust the set of alternatives than another. This corollary contradicts the implication of proportional pessimism that scientists of T_1 are most likely to exhaust the set of alternatives, and that scientists of T_3 are least likely to exhaust the set of alternatives.

How would pessimists respond to the foregoing two arguments for preferring the counter-induction to PUA? They might go on the offensive, raising the objection against the counter-induction that if current scientists can exhaust sets of alternatives, as the counter-induction asserts, current scientists should have alternatives to their theories in their hands, but they do not. For example, germ theorists should have an alternative to the germ theory in their hands, but they do not. Why do they not have an alternative?

Let me use an analogy to answer this question. You have gone to a supermarket to buy avocados, but you came back home empty-handed. Your spouse asks you why you came back home empty-handed. You answer that the supermarket was sold out of avocados, and that just because you came back home empty-handed, it does not follow that you did not have money to buy avocados. Similarly, germ theorists do not have an alternative to the germ theory in their hands. Pessimists of PUA ask counter-inductivists why germ theorists do not have an alternative. Counter-inductivists answer that the set of alternatives is empty, and that just because germ theorists do not have an alternative to it, it does not follow that they do not have the cognitive capacity to exhaust the set of alternatives to it.

Pessimists might object that since there were alternatives to T_1 and T_2 when T_1 and T_2 were accepted, there are alternatives to T_3 when T_3 is accepted. However, counter-inductivists would retort that since there were alternatives to T_1 and T_2 when T_1 and T_2 were accepted, there is no alternative to T_3 when T_3 is accepted, i.e., that since a set of unconceived alternatives had members when T_1 and T_2 were accepted, it does not have a member when T_3 is accepted. It became empty when scientists removed T_3 from it. As a consequence, it is of no use for pessimists to appeal to the history of science to argue that scientists of T_3 should have an alternative to T_3 in their hands.

7.5 The Counter-Induction Against PME

In this section, I construct a counter-induction against PME, and sketch the possible dialectics between pessimists of PME and counter-inductivists.

PME holds that since "Past scientists often faced misleading total evidence sets," it is likely that "Current scientists often face misleading total evidence sets" (Frost-Arnold, 2019, p. 912). PME does not make the agnostic claim that we do not know whether the evidence for a present theory is likely to be misleading, but rather the pessimistic claim that it is likely to be misleading. It follows that pessimists have the burden of establishing that PME is correct while the counter-induction against it is incorrect.

Proportional pessimism is also built into PME. According to PME, the fact that the evidence for T_1 was misleading indicated that the evidence for T_2 was misleading, and the fact that the evidence for T_2 was misleading indicated that the evidence for T_3 is misleading. Consequently, the fact that the evidence for T_1 and T_2 was misleading indicates more strongly that the evidence for T_3 is misleading than the fact that the evidence for T_1 was misleading indicated that the evidence for T_2 was misleading. In a nutshell, the evidence for T_1 was least likely to be misleading, and the evidence for T_3 is most likely to be misleading. Keep this implication of proportional pessimism in mind, as it will be invoked below.

The counter-induction against PME holds that because the evidence for past theories was misleading, it is likely that the evidence for contemporary theories is truth-leading. The premise of the counter-induction is the same as the premise of PME, but its conclusion is the opposite of the conclusion of PME. Pessimists of PME and counter-inductivists draw opposite conclusions from the same premise.

What if a current theory, say, the kinetic theory, is overthrown by an alternative? Pessimists of PME might excitedly say that they were right and counter-inductivists were wrong about the evidence for the kinetic theory, i.e., that the evidence for it was misleading as they thought and contrary to what counter-inductivists thought. However, counter-inductivists would retort that strictly speaking, the overthrowing of the kinetic theory is compatible with the counter-induction, pointing out that it did not assert in the first place that the evidence for the kinetic theory was *definitely* truth-leading, but rather that the evidence for it was *likely* to be truth-leading. They would add that the misleadingness of the evidence for the kinetic theory makes it more likely that the evidence for its successor is truth-leading. Therefore, the overthrowing of the kinetic theory would only deepen the disagreement between pessimists of PME and counter-inductivists.

How can we resolve the stalemate between pessimists of PME and counter-inductivists? In Sect. 7.3 above, I presented two arguments to resolve the stalemate between pessimists of CPI and counter-inductivists. The two arguments can also be invoked to resolve the stalemate between pessimists of PME and counter-inductions. Let me expound one by one how they can be used.

The first argument is that the evidence for new theories is more powerful than that for old theories. For example, Galileo observed Venus going through a phase, while Ptolemy did not. The evidence that Galileo had for the Copernican theory was more extensive than the evidence that Ptolemy had for the Ptolemaic theory. Therefore, counter-inductivists do not commit the gambler's fallacy when they maintain that unlike the evidence for past theories, the evidence for current theories is truth-leading.

The second argument is that PME is incoherent. Compare the evidence for T_1, the evidence for T_2, and the evidence for T_3. Proportional pessimism implies, as noted earlier in this section, that the evidence for T_1 is least likely to be misleading, and that the evidence for T_3 is most likely to be misleading. This implication of proportional pessimism contradicts the corollaries of both assumptions that the number of alternatives is finite and infinite, as demonstrated below.

Suppose first that a set of alternatives is finite. As more and more alternatives are eliminated from the set of alternatives, the number of remaining alternatives decreases, and the probabilities of the remaining alternatives increase. As a result, T_1 is least probable and T_3 is most probable. Since T_1 is least probable and T_3 is most probable, the evidence for T_1 is most probably misleading and the evidence for T_3 is least probably misleading. This corollary contradicts the implication of proportional pessimism that the evidence for T_1 was least probably misleading, and that the evidence for T_3 is most probably misleading.

Suppose now that a set of alternatives is infinite. No matter how finitely many alternatives were eliminated from the set of alternatives, there would still be infinitely members in the set of alternatives, and they would all be 0% probable. Accordingly, the evidence for T_1, the evidence for T_2, and the evidence for T_3 would all be equally misleading. This corollary contradicts the implication of proportional pessimism that the evidence for T_1 was least likely to be misleading, and that the evidence for T_3 is most likely to be misleading.

7.6 The English Notion

In the previous sections, I presented two arguments for preferring the three counter-inductions to the three pessimistic inductions. In this section, I reply to a possible move that pessimists might make regarding those arguments.

Pessimists might first argue that the two arguments are not rationally compelling. Suppose that there are one hundred alternatives to the kinetic theory. Under such a condition, the kinetic theory is likely to be false, even if the evidence for it is more powerful than that for the caloric theory, even if kinetic theorists are more likely to exhaust a set of alternatives than caloric theorists, and even if the evidence for the kinetic theory is more truth-leading than that for the caloric theory.

Next, pessimists might appeal to Bas van Fraassen (1989), who advances two competing notions of rationality: the Prussian and English notions. According to the former, "what is rational to believe is exactly what one is rationally compelled to believe" (van Fraassen, 1989, p. 171). According to the latter, by contrast, "what it is rational to believe includes anything that one is not rationally compelled to disbelieve" (van Fraassen, 1989, pp. 171–172). Van Fraassen (2007, p. 343, 2017, p. 99) invokes the latter to defend his position from his critics.

Pessimists might rely on the English notion to justify the inference that since we are not rationally compelled to prefer the three counter-inductions to the three pessimistic inductions, it is rational to prefer the three pessimistic inductions to the three counter-inductions, although it is also rational to prefer the three counter-inductions to the three pessimistic inductions. Thus, the English notion helps pessimists conclude that it is rational for them to believe, for example, that the kinetic theory is false, although it is also rational for counter-inductivists to believe that it is true.

To pessimists' surprise, however, counter-inductivists might welcome pessimists' conclusion above. Counter-inductivists might be content with pessimists' thesis that it is rational for counter-inductivists to prefer the three counter-inductions to the three pessimistic inductions, and counter-inductivists might not strive for the ambitious thesis that it is irrational for pessimists to prefer the three pessimistic inductions to the three counter-inductions. As a result, pessimists and counter-inductivists might stop debating and might not continue to argue against each other (Park, 2019b, p. 479).

So far in this section, I have assumed that the two arguments for preferring the three counter-inductions to the three pessimistic inductions are not rationally compelling, and hence that we are not rationally compelled to prefer the three counter-inductions to the three pessimistic inductions. Strictly speaking, however, this assumption is questionable. As I argued in the previous sections, the three pessimistic inductions are incoherent. Accordingly, we are rationally compelled to prefer the three counter-inductions to the three pessimistic inductions, and it is irrational to prefer the three pessimistic inductions to the three counter-inductions. For this reason, pessimists have the burden of refuting my arguments showing that the three pessimistic inductions are incoherent. Ideally, they would need to present some arguments in favor of their preference for the three pessimistic inductions rather than the three counter-inductions. As it stands, the English notion of rationality does not help pessimists.

7.7 Conclusion

Past scientists failed to obtain true theories, failed to conceive of alternatives to their theories, and failed to have truth-leading evidence for their theories. Pessimists and counter-inductivists have opposing views of what past scientists' failures indicate. For pessimists, they indicate current scientists' failures. For counter-inductivists, by contrast, they indicate current scientists' success. As a result, pessimists and counter-inductivists hold opposing views of the history of science. For pessimists, it is the graveyard of dead theories. For counter-inductivists, it is the birthplace of true theories. Let me conclude this chapter with a slogan: "Past science doesn't mirror current science."

References

Alai, M. (2017). Resisting the historical objections to realism: Is Doppelt's a viable solution? *Synthese, 194*(9), 3267–3290.

Alvaro, C. (2021). Atheism as an extreme rejection of rational evidence for the existence of God. *The Heythrop Journal, 62*(2), 1–16.

Chakravartty, A. (2008). What you don't know can't hurt you: Realism and the unconceived. *Philosophical Studies, 137*(1), 149–158.

Chang, H. (2011). The persistence of epistemic objects through scientific change. *Erkenntnis, 75*(3), 413–429.

Frost-Arnold, G. (2011). From the pessimistic induction to semantic antirealism. *Philosophy of Science, 78*(5), 1131–1142.

Frost-Arnold, G. (2019). How to be a historically motivated antirealist: The problem of misleading evidence. *Philosophy of Science, 86*(5), 906–917.

Graham, D. (2013). *Science before Socrates: Parmenides, Anaxagoras, and the new astronomy*. Oxford University Press.

Harker, D. (2015). *Creating scientific controversies: Uncertainty and bias in science and society*. Cambridge University Press.

Harrison, P. (2015). *The territories of science and religion*. University of Chicago Press.

Hempel, C. (1966). *Philosophy of natural science*. Prentice-Hall.

Laudan, L. (1977). *Progress and its problems: Towards a theory of scientific growth*. University of California Press.

Leplin, J. (1997). *A novel defense of scientific realism*. Oxford University Press.

Lipton, P. (2005). The truth about science. *Philosophical Transactions of the Royal Society B, 360*, 1259–1269.

Mizrahi, M. (2016). The history of science as a graveyard of theories: A philosophers' myth. *International Studies in Philosophy of Science, 30*(3), 263–278.

Müller, F. (2015). The pessimistic meta-induction: Obsolete through scientific progress? *International Studies in the Philosophy of Science, 29*(4), 393–412.

Musgrave, A. (1985). Realism vs. constructive empiricism. In P. Churchland & C. Hooker (Eds.), *Images of science: Essays on realism and empiricism*. The University of Chicago Press.

Park, S. (2016). Refutations of the two pessimistic inductions. *Philosophia, 44*(3), 835–844.

Park, S. (2017). The uniformity principle vs. the disuniformity principle. *Acta Analytica, 32*(2), 213–222.

Park, S. (2018a). The anti-induction for scientific realism. *Grazer Philosophische Studien, 95*(3), 329–342.

Park, S. (2018b). The grand pessimistic induction. *Review of Contemporary Philosophy, 17*, 7–19.

Park, S. (2019a). The absolute and relative pessimistic inductions. *Problemos, 95*, 94–104.

Park, S. (2019b). How to formulate scientific realism and antirealism. *Journal for General Philosophy of Science, 50*(4), 477–488.

Park, S. (2022). *Embracing scientific realism*. Springer.

Ruhmkorff, S. (2011). Some Difficulties for the Problem of Unconceived Alternatives. *Philosophy of Science, 78*(5), 875–886.

Ruhmkorff, S. (2013). Global and local pessimistic meta-inductions. *International Studies in the Philosophy of Science, 27*(4), 409–428.

Sorensen, R. (2013). Veridical idealizations. In M. Frappier, L. Meynell, & J. Brown (Eds.), *Thought experiments in philosophy, science, and the arts* (pp. 30–52). Routledge.

Stanford, K. (2006). *Exceeding our grasp: Science, history, and the problem of unconceived alternatives*. Oxford University Press.

Stengers, I. (2000). *The invention of modern science* (D. Smith, Trans.). University of Minnesota Press.

van Fraassen, B. (1989). *Laws and symmetry*. Oxford University Press.

van Fraassen, B. (2007). Reply: From a view of science to a new empiricism. In B. Monton (Ed.), *Images of empiricism: Essays on science and stances, with a reply from Bas van Fraassen* (pp. 337–383). Oxford University Press.

van Fraassen, B. (2017). Misdirection and misconception in the scientific realism debates. In E. Agazzi (Ed.), *Varieties of scientific realism: Objectivity and truth in science* (pp. 95–108). Springer.

Wray, B. (2013). Pessimistic induction and the exponential growth of science reassessed. *Synthese, 190*(18), 4321–4330.

Chapter 8
Stage Theory of Scientific Development

Abstract Stage theory holds that each field of science undergoes an unstable stage and then a stable stage. In the unstable stage, it is likely that scientists learn through trials and errors, and that an existing theory is false. In the stable stage, by contrast, it is likely that scientists no longer learn through trials and errors, and that an existing theory is true. According to stage theory, past scientists accepted a past theory because they were in an unstable stage, and current scientists accept a current theory because they are in a stable stage. This account is better than two pessimist accounts of why past scientists accepted a past theory and why current scientists accept a current theory.

Keywords Problem of misleading evidence · Problem of unconceived alternatives · Stage theory

8.1 Introduction

Scientists accept scientific theories. Past scientists accepted past theories, such as the geocentric system, the caloric theory, and the phlogiston theory. Current scientists accept current theories, such as the kinetic theory, evolutionary theory, and the oxygen theory. This paper concerns the following questions. Why did past scientists accept past theories? Why do current scientists accept current theories?

There are two competing answers to these questions in the literature. According to P. Kyle Stanford's (2006) problem of unconceived alternatives (PUA), past scientists accepted a past theory because they could not exhaust a set of alternatives to it, and current scientists similarly accept a current theory because they cannot exhaust a set of alternatives to it. According to Greg Frost-Arnold's (2019) problem of misleading evidence (PME), past scientists accepted a past theory because the evidence for it was misleading, and current scientists accept a current theory similarly because the evidence for it is misleading. These two accounts are pessimistic about scientists' cognitive ability to conceive of alternatives and the evidence for theories, respectively.

S. Park, *Induction, Science, and Morality*, Synthese Library 520,
https://doi.org/10.1007/978-3-032-00032-3_8

This chapter develops a realist alternative to the two pessimist accounts which I call "the stage account." It holds that past scientists accepted a past theory because they were in an unstable stage of scientific development, while current scientists accept a current theory because they are in a stable stage of scientific development. In an unstable stage, it is likely that scientists learn through trials and errors, and that an existing theory is false. An unstable stage is followed by a stable stage in which it is likely that scientists no longer learn through trials and errors, and that an existing theory is true.

This chapter has six sections. In Sect. 8.2, I unfold the PUA, PME, and stage accounts of why past scientists accepted a past theory. In Sect. 8.3, I unpack the three accounts of why current scientists accept a current theory, offer criticisms of the two pessimist accounts, and defend the stage account. In Sect. 8.4, I argue that a technological achievement sometimes comes after trials and errors, and that the existence of technological achievements supports stage theory and undermines pessimism. In Sect. 8.5, I run the same argument mutatis mutandis concerning predictive achievements. In Sect. 8.6, I construct a dilemma for pessimists of PME.

It is important to evaluate the three accounts of why scientists accept their theories because, depending on which of them you prefer, you may embrace or resist scientific realism, the view that "we can justifiably believe the hypothesis that successful theories are approximately true" (Lyons, 2005, p. 171). This chapter will be useful to those who wonder how science develops, and whether the development of science supports pessimism or realism.

8.2 Why Accepted Past Theories?

8.2.1 PUA

This subsection introduces the PUA account of why past scientists accepted a past theory and then introduces Frost-Arnold's (2019) objection to it.

PUA is intended to improve upon the classic pessimistic induction (CPI) propounded by Henri Poincaré (1905/1952, p. 160). CPI holds that most "of the past theories of science are already suspected of being false; there is presumably every reason to anticipate that current theories of science will suffer a similar fate" (Laudan, 1977, p. 126). Note that CPI ranges over theories, i.e., it is an inference from past theories to their current successor theories. Unlike CPI, PUA (Stanford, 2006, p. 20) ranges over scientists. It asserts that past scientists could not exhaust sets of alternatives, so it is likely that current scientists cannot exhaust sets of alternatives either.

PUA is intended to overcome a standard objection to CPI, namely that the evidence for present theories is better than that for past ones, and thus the downfall of present theories cannot be anticipated from that of past ones (Musgrave, 1985, p. 211; Leplin, 1997, p. 141). Stanford (2006, pp. 44–45) admits that the evidence for present theories is better than that for past theories. However, he argues that PUA ranges not over theories but over theorists, and that current theorists are not better able to exhaust sets of alternatives than past theorists. Past and current theorists are "creatures whose cognitive constitutions are not well suited to the task of exhausting the kinds of spaces of serious candidate theoretical explanations from which our scientific theories are drawn" (Stanford, 2006, p. 45).

PUA can be invoked to explain why past scientists accepted a past theory. According to PUA, past scientists accepted a past theory because they could not exhaust a set of alternatives to it. For example, humoral theorists accepted the humoral theory because they could not exhaust a set of alternatives which includes the miasma and germ theories. Similar accounts can be given of why past scientists accepted other past theories.

Frost-Arnold (2019, pp. 913–915) raises a few objections to the foregoing PUA account. The most forceful objection, I believe, is that some "historical examples involve conceived alternatives" (Frost-Arnold, 2019, p. 914). Frost-Arnold's examples are as follows. Ptolemy considered Aristarchus's heliocentric system but preferred the geocentric system to it. Caloric theorists also considered the kinetic theory but preferred the caloric theory to it. Such historical episodes are all anomalies to the PUA account. After all, they conflict with the PUA account that Ptolemy accepted the geocentric system because he could not conceive of an alternative to it, and that caloric scientists accepted the caloric theory because they could not conceive of an alternative to it.

8.2.2 PME

The premise of PME is that "Past scientists often faced misleading total evidence sets," and the conclusion of PME is that "Current scientists often face misleading total evidence sets" (Frost-Arnold, 2019, p. 912). Thus summarized, PME is distinct from PUA. PME capitalizes on the misleadingness of scientists' evidence, while PUA capitalizes on scientists' inability to conceive of alternatives. Therefore, PME is a new challenge to scientific realism.

Like PUA, PME is also intended to get around the aforementioned standard objection to CPI. Suppose that T_1, T_2, and T_3 are successive theories. T_1 and T_2 are obsolete theories, while T_3 is a current theory. The standard objection says that the evidence for T_3 is greater than that for T_1 and T_2, and thus the inference from the falsity of T_1 and T_2 to that of T_3 is fallacious. Pessimists of PME retort that although the evidence for T_2 was better than that for T_1, the former was misleading just as the latter was. It follows that although the evidence for T_3 is better than that for T_2, the former is misleading just as the latter was.

Now that we are clear about PME, we are ready to apply it to the issue of why past scientists accepted a past theory. According to PME, "the reason a particular past theory was accepted is that the evidence available at that earlier time was misleading" (Frost-Arnold, 2019, p. 912). This PME account is distinct from the PUA account. Moreover, PME explains not only why past scientists accepted a past theory but also why they rejected an alternative to a past theory. For example, Ptolemy considered the heliocentric system and rejected it because the evidence available to him "was misleading" (Frost-Arnold, 2019, p. 911).

8.2.3 Stage Theory

Stage theory holds that each field of science undergoes an unstable stage and then a stable stage. In this subsection, I flesh out stage theory, apply it to past scientists' acceptances of past theories, and then reply to a possible objection.

In the unstable stage, it is likely that scientists learn through trials and errors. They make the mistakes of taking false theories to be true and taking true theories to be false. They do not have sufficient negative knowledge (Bird, 2007, p. 108) that past theories were false. As a result, a scientific revolution is likely to occur, and thus an existing theory is likely to be false. For example, Ptolemaic scientists were in an unstable stage. They made the mistake of taking Aristarchus's heliocentric theory to be false, and it was likely that the Ptolemaic theory would be replaced with an alternative.

The unstable stage is followed by a stable stage. In the stable stage, it is likely that scientists no longer learn through trials and errors. They no longer make the mistake of taking false theories to be true and taking true theories to be false. They have sufficient negative knowledge that past theories were false. Accordingly, a scientific revolution is unlikely to occur, and thus an existing theory is likely to be true. For example, germ theorists are in a stable stage. They have the negative knowledge that the humoral and miasma theories were false, and it is unlikely that the germ theory would be replaced with an alternative.

The difference between the unstable and stable stages is a difference in degree, not in kind. The more scientists in a particular field make mistakes and experience failures, the closer they are to a stable stage, and the more probable it is that their theory is true. As scientists make more and more mistakes and failures, they go above a certain point that I call "a flipping point" (Park, 2017, p. 217). In this context, a flipping point is defined as the point at which a transition is made from an unstable stage to a stable stage. A field of science is in an unstable stage before it passes a flipping point, and it is in a stable stage after it passes the flipping point.

A flipping point is similar to what Richard Boyd calls "a take-off point." A take-off point is "a point in the development of the relevant scientific discipline at which the accepted background theories are sufficiently approximately true and comprehensive" (Boyd, 1980, p. 627). A difference between the two concepts is that a take-off point is defined in terms of the truth of background theories, while a flipping point is defined in terms of trials and errors. A similarity between them is that they share the view that as science develops, theories become more and more probable.

Different fields of current science are in different stages of scientific development. Germ theorists are in the stable stage. As noted earlier, it is unlikely that the germ theory will be replaced with an alternative. By contrast, suppose that current scientists just launched a new field of science. Such a field of science is in an unstable stage. The first theory in the new field of science needs to be replaced with a new one before the field of science enters a stable stage. At any rate, no stagist (theorist who espouses stage theory) would claim that all fields of current science are in stable stages.

Different fields of science require different numbers of mistakes and failures before they enter stable stages. Regarding combustion phenomena, there was only one scientific revolution, the transition from the phlogiston theory to the oxygen theory. By contrast, after Newton established the corpuscular paradigm about the nature of light, physicists rejected and accepted several theories of light (Fahrbach, 2011a, pp. 141–142). Different fields of science investigate different entities and use different methods. Consequently, different numbers of scientific revolutions are required for a field of science to move from an unstable stage to a stable stage.

How would pessimists respond to the concepts of unstable and stable stages? They would be pessimistic about the prospect of identifying a flipping point, the point at which a field of science moves from an unstable stage to a stable stage. It is unlikely that future philosophical research on diverse fields of science could tell us that physics requires, say, 64,378 failures and mistakes while psychology requires, say, 783,421 failures and mistakes. In short, it is unlikely that stagists will find the line between an unstable stage and a stable stage.

In response, I raise a similar objection to CPI, PUA, and PME. CPI, for example, infers the falsity of the oxygen theory from that of the phlogiston theory. This inference is susceptible to the charge of a hasty generalization. It is not clear exactly how many false predecessors the oxygen theory should have before we are entitled to infer its falsity. Pessimists' research project of determining the exact number of required false predecessors is no brighter or dimmer than stagists' research project of determining the exact number of required false predecessors. Therefore, pessimists and stagists have research projects of equal difficulty ahead of them.

More importantly, the difficulty of recognizing the line between an unstable stage and a stable stage should not be the grounds for dismissing the concepts of unstable and stable stages any more than the difficulty of finding the line between boys and men is the grounds for dismissing the concepts of boy and man. The vague terms "boy" and "man" are usable, provided that there are individuals who are clearly boys and individuals who are clearly men (van Fraassen, 1980, p. 16).

Relatedly, the absence of precise definitions is a prevalent phenomenon in philosophy of science. For example, we cannot draw the line between successful and unsuccessful scientific theories. Larry Laudan has given the most precise definition of success, according to which a theory is successful "so long as it has worked well, i.e., so long as it has functioned in a variety of explanatory contexts, has led to confirmed predictions and has been of broad explanatory scope" (Laudan, 1981, p. 23). In this admirably clear definition, I could not find a line between unsuccessful and successful theories. If we dismissed the notion of success on the grounds that we cannot see a line between unsuccessful and successful theories, the realism debate would collapse, for it surrounds the issue of whether "we can justifiably believe the hypothesis that successful theories are approximately true" (Lyons, 2005, p. 171).

The distinction between unsuccessful and successful theories is similar to the distinction between unstable and stable stages in a few respects. Just as the difference between unsuccessful and successful theories is a difference in degree rather than in kind, so the difference between unstable and stable stages is a difference in degree rather than in kind. Just as an unsuccessful theory becomes a successful one once the degree of success goes above a flipping point, so a field of science in an unstable stage enters a stable stage once the number of mistakes and failures goes above a flipping point. Just as the degree of success goes above a flipping point as it explains and predicts more and more phenomena, so the number of mistakes and failures goes above a flipping point as scientists make more and more mistakes and failures. Therefore, the two distinctions are in the same boat.

How does my distinction between unstable and stable stages differ from Thomas Kuhn's (1962/1970) distinction between immature and mature sciences? On Kuhn's account, an immature science is a field of science in which there is no established paradigm, and rival schools of thought have different views about fundamentals. An immature science becomes a mature science once one school of thought defeats other schools of thought and establishes a paradigm. A mature science undergoes "alternating phases of normal science and revolutions" (Bird, 2018). To put it differently, "the successive transition from one paradigm to another via revolution is the usual developmental pattern of mature science" (Kuhn, 1962/1970, p. 12).

Scientists learn through trials and errors in both immature and mature sciences. No theory is accepted in an immature science, and a theory accepted in a mature science. However, a theory accepted in a mature science is likely to be replaced by an alternative. It follows that both immature and mature sciences are in an unstable stage. The field of science devoted to investigating the nature of light was mature but was in an unstable stage when the several theories of light came and went after Newton.

According to Kuhn, scientific development consists of indefinite alternations of normal and revolutionary science. According to stagists, by contrast, the alternations of normal science and revolutionary science ultimately end in a stable stage. Kuhn and stagists share the picture of how a field of science developed in the past, viz., there were scientific upheavals in the field of science. However, they have opposite pictures of how the field of science will develop in the future. While Kuhn predicts that scientific upheavals are likely to occur, stagists predict that they are unlikely to occur in the field of science. As far as stagists are concerned, scientific revolutions are a means for the field of science to enter a stable stage.

Let me now apply stage theory to the following two examples from the history of science. Phlogiston theorists accepted the phlogiston theory because they were in an unstable stage. They were learning through trials and errors when they investigated combustion phenomena. In the process, they made the mistake of accepting a false theory. As another example, Ptolemaic scientists accepted the geocentric system rather than the heliocentric system because they were in an unstable stage. They were learning through trials and errors when they investigated the universe. In the process, they made the mistake of accepting a false theory rather than the true theory.

The foregoing stage explanations go hand in hand with the stage explanations that we offer in our daily lives. For example, we say that Suzie likes fairy tales because she is a child and children mostly like fairy tales. In this example, we explain the behavior of a person by saying that she is in a certain stage of development. I call such an explanation "a stage explanation." A stage explanation has the explanatory scheme that an object behaves in a certain manner because it is in a certain stage of development, and similar objects tend to behave similarly in that stage of development. This explanatory scheme is exemplified by the two stage explanations of why phlogiston scientists accepted the phlogiston theory and why Ptolemaic scientists accepted the geocentric system rather than the heliocentric system.

The stage account of past scientists' acceptance of a past theory is consistent with the PUA and PME accounts. The stage account that phlogiston theorists accepted the phlogiston theory because they were in an unstable stage is compatible with the PUA account that they accepted the phlogiston theory because they were not able to conceive of an alternative to it, and also with the PME account that they accepted the phlogiston theory because the evidence for it was misleading.

Pessimists would object that stage theory generates a circular explanation. According to stage theory, phlogiston theorists accepted the phlogiston theory because the field of science devoted to investigating combustion phenomena was in an unstable stage, and the field of science was in an unstable stage because phlogiston theorists accepted a false theory. It is circular to say that phlogiston theorists accepted a false theory because the field of science was in an unstable stage, and that the field of science was in an unstable stage because phlogiston theorists accepted a false theory.

Contrary to this objection, stage theory does not assert that the field of science was in an unstable stage because phlogiston theorists accepted a false theory; it rather asserts that the field of science was in an unstable stage because the phlogiston theory was the first one in the field of science devoted to investigating combustion phenomena. On stage theory, any field of science is in an unstable stage when its existing theory is the first one, and the first theory should be replaced with a new one before the field of science moves into a stable stage. It is improbable for a field of science to be in a stable stage without going through an unstable stage.

This view regarding the first theory in a field of science contrasts with the pessimist view on it. The pessimist view is silent on the fate of the first theory, i.e., it does not say anything about whether it will be replaced with a new theory. The first theory does not have a predecessor, and thus pessimists cannot say that since its predecessor was discarded, it will also be discarded. Only after the first theory is discarded can they predict that the second theory will be discarded (Park, 2018a, p. 13).

8.3 Why Accept Current Theories?

In the previous sections, I unfolded the PUA, PME, and stage explanations of why past scientists accepted past theories. In this section, I move on to different explananda. Why do current scientists accept current theories? For example, why do current scientists accept the oxygen theory? Pessimists and stagists give conflicting answers to this question.

Pessimists' answers are straightforward. Pessimists of PUA would say that current scientists accept a current theory because they cannot exhaust a set of alternatives to it. Pessimists of PME would say that current scientists accept a current theory because the evidence for it is misleading. On the two pessimist accounts, past and current scientists' cognitive capacities are in the same boat, and the evidence for past and current theories are also in the same boat. Accordingly, the explanantia for past scientists' acceptances of past theories are similar to those for current scientists' acceptances of current theories.

In contrast, stage theory invokes radically different explanantia to current scientists' acceptances of their theories. Specifically, it asserts that current scientists accept a current theory not because they are in an unstable stage but rather because they are in a stable stage. For example, germ theorists accept the germ theory not because they are in an unstable stage but rather because they are in a stable stage. The stage explanation implies that the germ theory is likely to be true. Consequently, the stage account of current scientists' acceptance of a current theory is a realist one, and it conflicts with the PUA and PME accounts.

What is the evidence for stagists' belief that a current field of science is in a stable stage? For example, what is the evidence for their belief that germ theorists are in a stable stage? The evidence for their belief is the same as pessimists' evidence for their belief that the current field of science was in an unstable stage. Pessimists claim that since the current field of science was in an unstable stage, it is and will be in an unstable stage. By contrast, stagists claim that since the current field of science was in an unstable stage, it is and will be in a stable stage.

What is the evidence for stagists' belief that an existing theory in a current field of science is likely to be true? It is the same as pessimists' evidence for their belief that an existing theory in a current field of science is likely to be false. Pessimists claim that since an existing theory in a past field of science was false, its successor in the current field of science is likely to be false. By contrast, stagists claim that since an existing theory in the current field of science was false, its successor in the current field of science is likely to be true. Pessimists and stagists draw opposite conclusions from the same premise.

Why do pessimists and stagists draw opposite conclusion from the same premise? They have opposing fundamental views about scientists. As far as stagists are concerned, scientists are agents who improve their abilities to obtain true theories, to conceive of alternatives, and to obtain truth-leading evidence. For this reason, stagists believe that since scientists had false theories in the past they now have true theories, that since scientists lacked the cognitive capacity to exhaust sets of alternatives in the past they now have the cognitive capacity, and that since scientists had misleading evidence in the past they now have truth-leading evidence. As far as pessimists are concerned, by contrast, scientists are not the kind of agents who improve those three abilities, or are not the kind of agents whose improved abilities make a difference to whether current theories are true or false, to whether current scientists can or cannot exhaust sets of alternatives, and to whether the evidence for current theories is misleading or truth-leading. Not surprisingly, pessimists draw pessimistic conclusions about current theories, current scientists' cognitive capacity, and about the evidence for current theories.

What if a scientific revolution occurs in a field of science that stagists take to be in a stable stage? Pessimists would contend that contrary to what stagists thought, the field of science was in an unstable stage. However, stagists would reply that strictly speaking, the scientific revolution is compatible with their old belief that the field of science was in an unstable stage. To say that a field of science is in a stable stage does not mean that an existing theory in the field of science is definitely true, but rather that it is likely to be true. Stagists would add that the scientific revolution will make it more likely that the field of science is in a stable stage. Therefore, the scientific revolution will only deepen the disagreement between pessimists and stagists.

The dispute between pessimists and stagists cannot be resolved merely by attaching the adjective "unconvincing" to each other's position. Pessimists would accuse stage theory of being unconvincing, saying that it sounds counterintuitive to them that since a field of science was in an unstable stage, it is and will be in a stable stage. In response, stagists would accuse the three pessimistic inductions of being unconvincing, saying that it equally sounds counterintuitive to them that since a field of science was in an unstable stage, it is and will be in an unstable stage. Therefore, an argument rather than labeling is required to resolve the dispute between pessimists and stagists.

Pessimists might ask stagists the following questions. Why draw the line in the present? Why not in the future? To state it differently, why say that because a field of science was in an unstable stage in the past, it is now in a stable stage? Why not say that because a field of science was in an unstable stage in the past, it is still in an unstable stage, but it will be in a stable stage in the future? It is arbitrary for stagists to assume that the field of science has already passed a flipping point.

Admittedly, the problem of arbitrariness plagues stage theory. However, similar problems plague CPI, PUA, and PME. Pessimists of CPI, for example, predict that the kinetic theory will be overturned in the future. But why in the future? Why not in the twentieth century? In other words, why not say that because the caloric theory was overturned, the kinetic theory was overturned in the twentieth century? Just as it is ad hoc for stagists to say that the kinetic theory is in a stable stage, so it is ad hoc for pessimists to say that the kinetic theory will be overthrown in the future.

In connection with this objection, I bring readers' attention to the fact that most scientific theories have been stable since the early twentieth century, as pointed out by historical optimists (Fahrbach, 2011a, 2011b; Park, 2011; Mizrahi, 2013, 2015, 2016). It follows that CPI would be reduced to absurdity if it asserted that since past theories were overturned before the twentieth century, their successors were overturned in the twentieth century. CPI has the chance to be correct only when it asserts that since past theories were overturned in the twentieth century, their successors will be overturned in the future. Therefore, pessimists have the burden of justifying their arbitrary choice of the future over the twentieth century.

8.4 Technological Achievements

In this section, I argue that scientists and engineers sometimes make achievements after they fail multiple times, and that the existence of such achievements supports stage theory and undermines pessimism.

Imagine that military engineers attempt to make a hypersonic cruise missile many times in vain. Under such circumstances, pessimists and stagists would make opposite predictions about whether military engineers will fail or succeed next time. Pessimists would infer that since military engineers were in an unstable stage in the past, they will continue to be in an unstable stage, which implies that it is likely that they will fail indefinitely. By contrast, stagists would infer that since military engineers were in an unstable stage in the past, they are now in a stable stage, which implies that it is likely that they will succeed next time. Military engineers actually succeed, as stagists predicted and contrary to what pessimists predicted.

How would pessimists respond to such an episode? They would say that a technological achievement sometimes comes after trials and errors, but that a theoretical achievement never comes after trials and errors. To state it otherwise, after engineers failed to make a new thing multiple times, they sometimes succeed in making a new thing. By contrast, even if scientists failed to obtain a true theory many times, they never succeed in obtaining a true theory. In short, pessimists would say that failure is sometimes the mother of success in technological development, but never in theoretical development (Park, 2018b, p. 333).

How are we to evaluate the foregoing pessimists' position? It embeds a double standard for technological and theoretical developments. It is ad hoc to say that an achievement sometimes comes after trials and errors in technological development, but never in theoretical development. Compare pessimists' position with stagists' position that an achievement sometimes come after trials and errors both in technological and theoretical developments. No double standard is involved in stagists' position, and hence stagists' position is more elegant than pessimists' position (Park, 2018b, p. 334).

Relatedly, we can raise the following objection to pessimists. Why is it that engineers sometimes make an achievement after trials and errors in technological development, but that scientists can never make an achievement after trials and errors in theoretical development? What is the relevant difference between technological and theoretical developments that would justify pessimists in asserting that failure is sometimes the mother of success in technological development but never in theoretical development? No answer to this question is forthcoming (Park, 2018b, p. 334).

Pessimists might complain that I have unfairly exaggerated their position. Their position is not that a theoretical achievement never comes after trials and errors, but rather that we do not know whether a theoretical achievement sometimes comes after trials and errors. On this account, even if we have a true theory in our hands, we do not know that we do. The justification for this position is that earlier theories were false, past scientists could not exhaust sets of alternatives, and the evidence for earlier theories was misleading.

This intriguing position is, unfortunately, incompatible with CPI, PUA, and PME. Pessimists do not make the skeptical claims that we do not know whether current theories are true or false, whether current scientists can or cannot exhaust sets of alternatives, and whether the evidence for current theories is truth-leading or misleading. They rather make the pessimistic claims that present theories are false, current scientists cannot exhaust sets of alternatives, and that the evidence for present theories is misleading. It is Humean skeptics, not pessimists, who make skeptical claims about current science, and Humean skepticism conflicts with a pessimistic induction (Park, 2019, pp. 142–144).

Suppose that inductivists observed that achievements sometimes came after trials and errors in an old domain, but they have not yet observed that achievements sometimes come after trials and errors in a new domain. Should they infer that since achievements sometimes came after trials and errors in the old domain, achievements also sometimes come after trials and errors in the new domain? My answer is that it depends. If they have good reason for thinking that the old and new domains are similar to each other, they should, as inductivists, make the inductive inference. By contrast, if they have good reason for thinking that the old and new domains are dissimilar to each other, they should not make the inductive inference. To put it conversely, their refusal to make the inductive inference gives rise to the burden of pointing out a relevant difference between the old and new domains. They cannot invoke Humean skepticism to justify their refusal to make the inductive inference. Specifically, they cannot say that since induction is not reliable, they refrain from making the inductive inference. After all, it would be self-defeating for inductivists to say so.

Now, what would be pessimists' justification for their refusal to make the inductive inference that since achievements sometimes come after trials and errors in technological development, achievements also sometimes come after trials and errors in theoretical development? It would be self-defeating for them to invoke Humean skepticism to justify their refusal to make the inductive inference. Specifically, it would be self-defeating for them to say that since induction is not reliable, our observation that achievements sometimes come after trials and errors in technological development does not justify the hypothesis that achievements also sometimes come after trials and errors in theoretical development. After all, they are inductivists making inductive inferences from past science to current science. If they were skeptics about induction, they could not make those inductive inferences and they would cease to be pessimists. Therefore, they have the burden of illuminating a relevant difference between technological and theoretical developments that would justify their refusal to make the inductive inference from the occasional achievements in technological development to those in theoretical development.

Pessimists might complain that it is unfair for me to talk about technological achievements in the context of the debate between pessimists and stagists. The issue between them is not about whether scientists have made technological achievements, but rather about whether current scientists have made theoretical achievements, whether current scientists can exhaust sets of alternatives, and whether the evidence for current scientific theories is truth-leading or misleading. Technological achievements are red herrings in the debate between pessimists and stagists.

However, realists would make a similar complaint to pessimists. Realists are interested in whether current theories are true or false, but not in whether past theories are true or false. They are interested in whether current scientists can or cannot exhaust sets of alternatives, but not in whether past scientists could or could not exhaust sets of alternatives. They are interested in whether the evidence for current theories is truth-leading or misleading, but not in whether the evidence for past theories was truth-leading or misleading. The truth-values of past theories, past scientists' cognitive capacity, and the evidence for past theories are all red herrings for the aforementioned issues that realists are interested in.

8.5 Predictive Achievements

In this section, I argue that scientists sometimes succeed in making a true prediction after they fail multiple times to make a true prediction, and that the existence of such achievements supports stage theory and undermines pessimism.

Scientists sometimes make a true prediction after going through trials and errors. To put it differently, an achievement sometimes comes after trials and errors in predictional development, a series of events in which scientists make multiple attempts to make a true prediction. How would pessimists respond to the existence of predictive achievements? They would admit that achievements sometimes come after trials and errors in predictional development. However, they would insist that achievements never come after trials and errors in theoretical development.

How are we to make of the foregoing pessimists' position? It embeds a double standard for predictional and theoretical developments. It is ad hoc to say that failure is sometimes the mother of success in predictional development, but never in theoretical development. The fact that achievements sometimes come after trials and errors in predictional development is an inductive rationale for thinking that achievements also sometimes come after trials and errors in theoretical development. Pessimists' refusal to make the inductive inference from occasional achievements in predictional development to those in theoretical development gives rise to the burden of specifying a relevant difference between predictional and theoretical developments.

There is a significant difference between predictive and technological achievements. A predictive achievement is a cognitive achievement, whereas a technological achievement is an interventive achievement. The cognitive achievement is a positive result of observing nature. Observing nature is a passive affair in that nature does something to scientists when they observe nature. By contrast, the interventive achievement is a positive result of intervening in nature. Intervening in nature is an active affair in that scientists do something to nature when they intervene in nature. Therefore, I am not invoking similar achievements but rather radically different achievements to argue that pessimists' position is ad hoc.

How do theoretical, predictional, and theoretical developments relate to one another? Predictional and theoretical developments are both cognitive developments in that they are both developments toward the production of new beliefs. By contrast, technological development is an interventive development, a development toward the production of new things. Technological development is different in kind from predictional and theoretical developments. Consequently, theoretical development is more similar to predictional development than to technological development.

This conclusion has the following two noteworthy implications against pessimists. (1) It is more ad hoc for pessimists to say that failure is sometimes the mother of success in predictional development, but never in theoretical development, than to say that failure is sometimes the mother of success in technological development, but never in theoretical development. (2) The fact that failure is sometimes the mother of success in predictional development is a stronger inductive rationale for thinking that failure is also sometimes the mother of success in theoretical development than is the fact that failure is sometimes the mother of success in technological development.

In Sect. 8.4 above, I claimed that it is ad hoc for pessimists to say that achievements sometimes come after trials and errors in technological development, but never in theoretical development. Earlier in this section, I claimed that it is more ad hoc for pessimists to say that achievements sometimes come after trials and errors in predictional development, but never in theoretical development. I add here that it is all the more ad hoc for pessimists to say that achievements sometimes come after trials and errors both in technological and predictional developments, but never in theoretical development.

Relatedly, in Sect. 8.4 above, I claimed that the fact that achievements sometimes come after trials and errors in technological development is an inductive rationale for thinking that achievements also sometimes come after trials and errors in theoretical development. Earlier in this section, I claimed that the fact that achievements sometimes come after trials and errors in predictional development is a *stronger* inductive rationale for thinking that achievements sometimes come after trials and errors in theoretical development. I add here that the fact that achievements sometimes come after trials and errors in technological and predictional developments is a *much stronger* inductive rationale for thinking that achievements sometimes come after trials and errors in theoretical development.

Pessimists might reply to this challenge as follows. Both technological and predictional developments concern observational entities, while theoretical development concerns theoretical entities. This difference justifies their ad hoc position that achievements sometimes come after trials and errors in technological and predictional developments, but never in theoretical development. Pessimists might flesh out their position as follows. Observational and theoretical entities fall on a continuum. A tree is clearly an observational entity while an electron is clearly a theoretical entity. There are diverse objects between them, for example viruses and molecules. As we investigate smaller and smaller objects, we go above a flipping point. Before we go above the flipping point, achievements sometimes come after trials and errors. After we go above the flipping point, by contrast, achievements never come after trials and errors.

This reply is saddled with the burden of exposing a relevant difference between observational and theoretical entities that would entitle them to say that achievements sometimes come after trials and errors in scientists' research about observational entities but achievements never come after trials and errors in scientists' research about theoretical entities (Park, 2018b, p. 335). What is so special about theoretical entities that makes pessimists think that scientists never make achievements in their research about them, although they sometimes make achievements in their research about observational entities?

Admittedly, as Bas van Fraassen argues, the terms "observables" and "unobservables" are usable insofar as there are clear cases of observables and unobservables, even if we cannot make a strict distinction between them (van Fraassen, 1980, p. 16). However, it is not clear what the justification would be for saying that technological and predictional developments are clear cases in which achievements sometimes come after trials and errors, and that theoretical development is a clear case in which achievements never come after trials and errors.

The prospect of satisfactorily meeting this objection is dim. Consider that there is a difference between technological and predictional developments, which is that scientists do something to nature in technological development while nature does something to scientists in predictional development. Despite this radical difference, however, achievements sometimes come after trials and errors both in technological and predictional developments. The radical difference does not justify the ad hoc position that achievements sometimes come after trials and errors in technological development, but not in predictional development. Nor does the difference justify the refusal to make an inductive inference from occasional achievements in technological development to those in predictional development.

By parity of reasoning, the difference that both technological and predictional developments concern observational entities while theoretical development concerns theoretical entities does not justify pessimists' ad hoc position that achievements sometimes come after trials and errors both in technological and predictive achievements but never in theoretical development. Nor does the difference justify the refusal to make an inductive inference from occasional achievements in technological and predictional achievements to those in theoretical development.

8.6 The Dilemma

As we have seen in Sect. 8.2.2 above, Frost-Arnold argues that PME can explain more phenomena than PUA. It appears that pessimists of PME aim to construct a better pessimistic induction than those that came before. A perverse question arises for pessimists of PME. Are they in an unstable stage or in a stable stage? Both answers are problematic, as shown below.

On the one hand, pessimists of PME might say that they are in an unstable stage. Like pessimists of CPI and PUA, they learn through trials and errors with the aim of developing a correct pessimistic induction. As noted in Sect. 8.2.1 above, CPI is vulnerable to the objection that the evidence for current theories is more powerful than that for past theories, and PUA is susceptible to Frost-Arnold's objection that it has anomalies. Since pessimists of CPI and PUA were in an unstable stage, pessimists of PME are also in an unstable stage. However, an implication of this conclusion is that PME is likely to be incorrect.

On the other hand, pessimists of PME might say that they are in a stable stage, and that PME is likely to be correct. PUA improves upon CPI, and PME improves upon PUA. Accordingly, PME is in a stable stage although CPI and PUA were in an unstable stage. However, this inference is vulnerable to the objections surrounding the concepts of stable and unstable stages that we discussed in Sect. 8.2.3 above. How many failures are required for pessimists to come up with a correct pessimistic induction? How do we know that pessimists of PME have passed the flipping point between an unstable stage and a stable stage?

More importantly, it would be a double standard for pessimists of PME to say that since PME improves upon its predecessors it is in a stable stage, but although current theories, current scientists, and the evidence for current theories improve upon their respective predecessors, they are in an unstable stage. Scientists would say, to the surprise of pessimists of PME, that although PME improves upon its predecessors it is in an unstable stage, but that since current theories, current scientists, and the evidence for current theories improve upon their respective predecessors, they are in a stable stage. In short, scientists would apply the opposite double standard to pessimists of PME.

To sum up, pessimists of PME are in a dilemma. If they say that they are in an unstable stage, they have to admit that PME is likely to be incorrect. If they say that they are in a stable stage, they have the burden of explaining why their improvement makes a difference to whether PME is in an unstable or in a stable stage, but why scientists' improvement does not make a difference to whether current theories are true or false.

Pessimists of PME might object that contrary to what the dilemma assumes, PME is not a successor to PUA. Nor is PUA a successor to CPI. PME is not intended to replace PUA. Nor is PUA intended to replace CPI. CPI, PUA, and PME are rather intended to reveal different aspects of the development from past to current science. CPI is intended to describe the unchanging truth-values of theories; PUA is intended to depict scientists' unchanging cognitive capacity to exhaust sets of alternatives;

PME is intended to reflect the unchanging misleadingness of the evidence for theories. CPI, PUA, and PME are all first pessimistic inductions in their own domains, and none of them has a predecessor.

This suggestion invites the following objection. If CPI, PUA, and PME are all first pessimistic inductions in their own domains, they are all in unstable stages. For example, if PME is the first pessimistic induction in its domain, it needs to be replaced with a new pessimistic induction before a correct pessimistic induction emerges in the domain. The same holds for CPI and PUA. In short, the suggestion that none of the three pessimistic inductions has a predecessor falls prey to the first horn of the aforementioned dilemma.

8.7 Conclusion

Scientists accepted a past theory, according to PUA, because they could not exhaust a set of alternatives to it; according to PME, because the evidence for it was misleading; according to stage theory, because scientists were in an unstable stage. Scientists accept a current theory, according to PUA, because they cannot exhaust a set of alternatives to it; according to PME, because the evidence for it is misleading; according to stage theory, because scientists are in a stable stage.

Pessimists believe that achievements sometimes come after trials and errors in technological and predictional developments but never in theoretical development. By contrast, stagists believe that achievements sometimes come after trials and errors in technological, predictive, and theoretical developments. Stagists' position is more elegant than pessimists' position. Pessimists of PME face the dilemma that either PME is likely to be incorrect or they hold a double standard for PME and science. Let me sum up this chapter with a slogan: "Science moves toward stability."

References

Bird, A. (2007). What is scientific progress? *Noûs, 41*(1), 64–89.

Bird, A. (2018). Thomas Kuhn. In E. N. Zalta (Ed.), *The Stanford encyclopedia of philosophy*. Retrieved from https://plato.stanford.edu/archives/win2018/entries/thomas-kuhn/

Boyd, R. (1980). Scientific realism and naturalistic epistemology. *PSA Proceedings of the biennial meeting of the philosophy of science association, 1980*(2), 613–662.

Fahrbach, L. (2011a). How the growth of science ends theory change. *Synthese, 180*(2), 139–155.

Fahrbach, L. (2011b). Theory change and degrees of success. *Philosophy of Science, 78*(5), 1283–1292.

Frost-Arnold, G. (2019). How to be a historically motivated antirealist: The problem of misleading evidence. *Philosophy of Science, 86*(5), 906–917.

Kuhn, T. (1962/1970). *The structure of scientific revolutions*. University of Chicago Press.

Laudan, L. (1977). *Progress and its problems: Towards a theory of scientific growth*. University of California Press.

Laudan, L. (1981). A confutation of convergent realism. *Philosophy of Science, 48*(1), 19–49.

Leplin, J. (1997). *A novel defense of scientific realism*. Oxford University Press.
Lyons, T. (2005). Toward a purely axiological scientific realism. *Erkenntnis, 63*(2), 167–204.
Mizrahi, M. (2013). The pessimistic induction: A bad argument gone too far. *Synthese, 190*(15), 3209–3226.
Mizrahi, M. (2015). Historical inductions: New cherries, same old cherry-picking. *International Studies in the Philosophy of Science, 29*(2), 129–148.
Mizrahi, M. (2016). The history of science as a graveyard of theories: A philosophers' myth. *International Studies in Philosophy of Science, 30*(3), 263–278.
Musgrave, A. (1985). Realism vs. constructive empiricism. In P. Churchland & C. Hooker (Eds.), *Images of science: Essays on realism and empiricism*. The University of Chicago Press.
Park, S. (2011). A confutation of the pessimistic induction. *Journal for General Philosophy of Science, 42*(1), 75–84.
Park, S. (2017). The uniformity principle vs. the disuniformity principle. *Acta Analytica, 32*(2), 213–222.
Park, S. (2018a). The grand pessimistic induction. *Review of Contemporary Philosophy, 17*, 7–19.
Park, S. (2018b). The anti-induction for scientific realism. *Grazer Philosophische Studien, 95*(3), 329–342.
Park, S. (2019). New objections to the problem of unconceived alternatives. *Filosofia Unisinos, 20*(2), 138–145.
Poincaré, H. (1905/1952). *Science and hypothesis*. Dover.
Stanford, K. (2006). *Exceeding our grasp: Science, history, and the problem of unconceived alternatives*. Oxford University Press.
van Fraassen, B. (1980). *The scientific image*. Oxford University Press.

Chapter 9
The Evolutionary Debunking Argument Against Scientific Realism

Abstract I raise three objections to de Ray's evolutionary debunking argument against scientific realism, and then sketch two new enhanced evolutionary debunking arguments that evolutionary debunkers might construct to overcome my three objections to de Ray's evolutionary debunking argument. I argue that any evolutionary debunking argument against scientific realism is self-defeating, and that any evolutionary debunking argument, whether it is directed at scientific realism or other philosophical views, such as mathematical realism and moral realism, is arbitrary.

Keywords Evolutionary debunking · Explainability principle · Scientific realism · Simplicity principle

9.1 Introduction

Christophe de Ray (2022) constructs an evolutionary[1] debunking argument against scientific realism. This argument against scientific realism is a novel one. Its magnitude is comparable to those of the major arguments against scientific realism: Larry Laudan's (1977, p. 126) pessimistic induction, Bas van Fraassen's (1989) objections to inference to the best explanation (IBE), and P. Kyle Stanford's (2006) problem of unconceived alternatives. Scientific realists should take all these objections seriously and should critically respond to them.

This chapter aims to criticize de Ray's debunking argument and to show that it is in principle impossible to discredit scientific realism with a debunking argument. In Sect. 9.2, I explicate de Ray's debunking argument in detail. In Sect. 9.3, I raise three objections to it. In Sect. 9.4, I reply to de Ray's objections to scientific realism and the no-miracles argument (Putnam, 1975, p. 73). In Sect. 9.5, I sketch and criticize two new enhanced debunking arguments that debunkers might construct to

[1] I drop "evolutionary" from now on.

S. Park, *Induction, Science, and Morality*, Synthese Library 520,
https://doi.org/10.1007/978-3-032-00032-3_9

overcome my three objections to de Ray's debunking argument. In Sect. 9.6, I argue that any evolutionary debunking argument, whether it is directed at scientific realism or other philosophical views, such as mathematical realism and moral realism, is arbitrary. In Sect. 9.7, I reply to a possible move that debunkers might make.

This chapter will be useful to those who wonder whether there can be a convincing debunking argument against scientific realism, and to those who wonder whether there is a common problem with all the debunking arguments against scientific realism, moral realism, mathematical realism, and so forth.

9.2 The Explainability Principle

In this section, I expound de Ray's debunking argument against scientific realism.

Let me first clarify a key term. We have the psychological inclination to believe that "facts generally have explanations for their obtaining" (de Ray, 2022, p. 970). It is due to this psychological inclination that we believe that there are explanations, for example, for the observational fact that a rock falls down and for the theoretical fact that a quark has mass. The general belief that most facts have explanations will be henceforth be referred to as "the explainability principle," as it was so referred to in Chaps. 2, 4, and 5.

Why do we have the psychological inclination to form the explainability principle? In other words, why are we inclined to believe that most facts have explanations? An evolutionary answer goes as follows. Our distant ancestors acquired it as a result of variations. Those who had it were more likely to survive and reproduce than those who did not. It was inherited by next generations. Over many generations, those who had it persisted while those who did not perished. As a result, we have it. In short, we have it because it was an adaptive trait.

Debunkers and defenders alike endorse the foregoing explanation of why we have the psychological inclination, and thus they agree that the psychological inclination was an adaptive trait. However, they disagree on whether it tracked truths. Defenders affirm, while debunkers deny, that it tracked truths. In consequence, defenders say, while debunkers do not, that we have the psychological inclination because it tracked truths. This explanation indicates that the psychological inclination was reliable, and thus that the explainability principle which the psychological inclination produces is justified.

Debunkers' explanation and defenders' explanation compete with each other. Debunkers' explanation holds that we have the psychological inclination because it was fitness-conducive, while defenders' explanation holds that we have the psychological inclination because it was fitness-conducive and it was fitness-conducive because it tracked truths. On what grounds can debunkers choose their explanation over defenders' explanation?

De Ray speculates that even if the psychological inclination tracked truths, it might be a lucky accident that we have the psychological inclination. He writes that "it would have taken an epistemically problematic 'lucky accident' for the evolutionary process to have given us sound innate metaphysical intuitions" (de Ray, 2022, p. 986). In other words, even if defenders' explanans that the psychological inclination was reliable were true, there might be no explanatory connection between the explanans and the explanandum that we have the psychological inclination. In short, the explanandum could have occurred independently of the explanans.

In addition, de Ray speculates that defenders' explanans might be false. On his account, the psychological inclination could be an adaptive trait even if it was not reliable. Empiricists, those who do not believe the explainability principle, could be "reproductively successful" (de Ray, 2022, p. 976). It is questionable whether it would have been beneficial for hominids to believe that theoretical facts, such as the fact that a quark has mass, have explanations (de Ray, 2022, p. 977). In short, "high reproductive fitness is perfectly compatible with having unsound innate metaphysical intuitions" (de Ray, 2022, p. 977). To state it differently, even if the psychological inclination were unreliable, we could be reproductively successful, i.e., even if defenders' explanans were false, the explanandum that we have the psychological inclination could be true.

After establishing the possibility that the psychological inclination could be an adaptive trait even if it was not reliable, de Ray infers that its reliability would not explain why we have it. To put it in de Ray's terms, since "the innate metaphysical intuitions we actually have could not have been selectively retained *by virtue* of being sound," "the soundness of our innate metaphysical intuitions would not at all explain our reliance on said intuitions" (de Ray, 2022, p. 977).

After inferring that the reliability of the psychological inclination would not explain why we have the psychological inclination, de Ray concludes that we ought to distrust the psychological inclination, i.e., we "ought to distrust (i.e. refrain from relying on) innate metaphysical intuition" (de Ray, 2022, p. 974). To say that we ought to distrust the psychological inclination implies that we are not justified in believing the explainability principle. Given that the explainability principle is required for our best scientific theories to be justified, we are not justified in believing them. This corollary clashes with scientific realism.

De Ray's debunking argument instantiates the argument scheme that "if you believe that the origins of the belief-forming method M on which you rely for some of your beliefs are such that the reliability of M would not at all explain your reliance on M, your continued reliance on M is unjustified" (de Ray, 2022, p. 974). This argument scheme can also be illustrated by de Ray's (2022, p. 973) example of the Bible. Your first-order religious beliefs result from the Bible. However, you newly acquire the second-order belief that you have first-order religious beliefs because your ancestors have chosen the Bible over the Quran for a political reason as opposed to an epistemic reason. You now have the second-order belief that the origin of the Bible is such that the reliability of the Bible would not explain why you have first-order religious beliefs. Under such circumstances, your first-order religious beliefs are not justified.

To sum up, we might accidently have the psychological inclination even if it was reliable, i.e., there might be no explanatory connection between defenders' explanans and the explanandum. In addition, the psychological inclination could be an adaptive trait even if it was not reliable, i.e., defenders' explanans might be false even if the explanandum is true. Therefore, the reliability of the psychological inclination cannot explain why we have the psychological inclination, i.e., defenders' explanans cannot explain the explanandum. This conclusion has the implication that the explainability principle and our best theories are not justified.

9.3 The Three Objections

9.3.1 *The Mere Possibilities*

In this subsection, I criticize de Ray's following argument. We might accidently have the psychological inclination even if it was reliable. It could be an adaptive trait even if it was not reliable. Therefore, the reliability of it would not explain why we have it.

The two premises are agreeable. After all, they are merely possibility statements. The refutation of the possibility premises requires the proof that it is 100% probable that we have the psychological inclination because it was reliable, i.e., that it is 100% probable that defenders' explanation is true. No one can provide such a proof. In general, if some imaginative thinkers claim that something is possible, we should agree with them except in cases where a contradiction is committed. For example, suppose that Cartesian skeptics claim that we might be brains in vats. We should agree with them, given that it is not a contradiction to say that we might be brains in vats. The same holds for de Ray. We should agree with him that we might accidently have the psychological inclination even if it was reliable, and that it could be an adaptive trait even if it was not reliable.

How about the inference from these two premises to the conclusion that the reliability of the psychological inclination would not explain why we have it? In my view, it is fallacious. The conclusion is not a possibility statement but a definite statement. It does not make the possibility claim that the reliability of the psychological inclination might not explain why we have it, but rather makes the definite claim that the reliability of it would not explain why we have it. In general, a definite statement does not even probably follow from possibility statements. Therefore, it is one thing that we might accidently have the psychological inclination even if it was reliable, and that it could be an adaptive trait even if it was not reliable. It is quite another that the reliability of it would not explain why we have it.

De Ray's inference from the two possibility premises to the definite conclusion can be reduced to absurdity. According to evolutionary theory, our distant ancestors were fish. This evolutionary hypothesis explains why human embryos have gills. The following objection can be raised to this evolutionary hypothesis. Human embryos might accidently have gills even if our distant ancestors were fish. In addition, human embryos can have gills even if our distant ancestors were not fish. Therefore, the evolutionary hypothesis that our distant ancestors were fish does not explain why human embryos can have gills. Similar objections can be raised to all the other hypotheses constituting evolutionary theory (and other scientific theories). In consequence, evolutionary theory would explain nothing. Since it would explain nothing, it would be unwarranted, and the debunking argument, which relies on the unwarranted theory, would be unconvincing. This absurd consequence follows from de Ray's inference from the two possibility premises to the definite conclusion.

The way de Ray argues against defenders' hypothesis that the psychological inclination was reliable is different from the way scientists argue against their opponents' scientific theory. Scientists do not appeal to the mere possibility that there might be no explanatory connection between their opponents' scientific theory and its explananda, and to the mere possibility that their opponents' scientific theory might be false even if its explananda are true; they rather argue that their opponents' scientific theory does not have virtues, such as accuracy, simplicity, coherence, and fruitfulness, or that their theory is more virtuous than their opponents' theory.

De Ray would be entitled to say that the reliability of the psychological inclination would not explain why we have it, provided that he shows that defenders' explanation that we have it because it was reliable does not meet the requirements that the major theories of scientific explanation claim that scientific explanations should meet. Different requirements are proposed by the deductive-nomological model (Hempel, 1966), the unification theory (Friedman, 1974), the causal theory (Salmon, 1978), and the contextual theory (van Fraassen, 1980). To state it briefly, they propose, respectively, that one of the explanantia should be a law of nature, that an explanans should unify phenomena, that an explanans should be a cause while an explanandum should be an effect, and that there should be a context in which an explanans is an appropriate answer to a why-question.

Suppose that de Ray establishes the following four theses. (1) There is no law of nature in defenders' explanation. (2) The reliability of the psychological inclination does not unify diverse phenomena. (3) It is false that the reliability of the psychological inclination is a cause, and that our retention of it is an effect. (4) There is no context in which it is legitimate to say that we have the psychological inclination because it was reliable. Establishing these four theses would make it plausible that the reliability of the psychological inclination would not explain why we have it.

None of the foregoing four theories of scientific explanation requires that there should be no possibility at all that there might be no explanatory connection between an explanans and an explanandum, and that the explanans might be false even if the explanandum is true. This observation of the four theories of scientific explanation creates the burden for de Ray to specify under what model of scientific explanation he operates when he argues that the reliability of the psychological inclination would not explain why we have it because we might accidently have it even if it was reliable and it could be an adaptive trait even if it was not reliable.

De Ray's conclusion that the reliability of the psychological inclination would not explain why we have it requires not the two possibility premises but rather the two probabilistic premises that it is likely that we accidently have it even if it was reliable, and that it was an adaptive trait even if it was not reliable. If de Ray establishes these two probabilistic premises, defenders will happily give up their explanation that we have the psychological inclination because it was reliable. To use an analogy, if creationists establish the probabilistic statements that it is likely that human embryos accidently have gills even if our distant ancestors were fish, and that human embryos have gills even if our distant ancestors were not fish, evolutionists would happily give up their explanation that human embryos have gills because our distant ancestors were fish.

How could de Ray establish the probabilistic premise that it is likely that the psychological inclination was an adaptive trait even if it was not reliable? He could do it by constructing the enumerative induction that since most of the facts that we have encountered did not have explanations, most facts do not have explanations. Once he established the enumerative induction, it would be likely that the psychological inclination was an adaptive trait even if it was not reliable.

However, it would be difficult to establish the enumerative induction. Suppose that de Ray presents some facts that he takes to be brute to justify the enumerative induction. Defenders would object that the putative brute facts, although we do not have their explanations in our hands, do have explanations. Further scientific research on the alleged brute facts will produce explanations for them. Defenders might even construct an inductive argument for this objection, citing the cases in the history of science in which some facts that initially appeared to be brute turned out to have explanations.

Let me wrap up this subsection with the thought on Cartesian skeptics' conclusion that physical beliefs, such as the belief that we have two hands, are not justified. This conclusion requires not the mere possibility that we might be brains in vats, but rather the high probability that we are brains in vats. After all, epistemologists do not claim that a justified belief is 100% probable but rather highly probable (Goldman & Beddor, 2021). Therefore, physical beliefs can be shown to be unjustified not by the mere possibility that they might be false, but rather by a high probability that they are false. A mere possibility does not make a physical belief unjustified any more than it makes an explanation inadequate.

9.3.2 The Enumerative Induction

In this subsection, I argue that even if the reliability of the psychological inclination would not explain why we have it, we can nonetheless establish the reliability of it.

As noted in Sect. 9.2 above, de Ray argues that since the reliability of the psychological inclination would not explain why we have it, we should distrust it, i.e., we "ought to distrust (i.e. refrain from relying on) innate metaphysical intuition" (de Ray, 2022, p. 974). He writes that "if you believe that the origins of the belief-forming method M on which you rely for some of your beliefs are such that the reliability of M would not at all explain your reliance on M, your continued reliance on M is unjustified" (de Ray, 2022, p. 974). It follows that the explainability principle is not justified.

What are we to make of de Ray's inference? In my view, it is fallacious. Imagine that a primitive person accidently picks up a telescope in a jungle. He sees what appears to be a bird in a distant tree through the telescope. He walks to the tree and sees a bird with his naked eyes, thereby establishing the reliability of the telescope. Under such circumstances, his first-order beliefs that he forms with the use of the telescope are justified. They are justified even if he has the second-order believe that the origin of the telescope is such that its reliability would not explain why he relies on it. After all, when asked why he relies on it, he can provide a story of how he calibrated it.

This imaginary episode has the following philosophical lesson. Our first-order beliefs are justified or not, not depending on whether we have the second-order belief or not that the origin of the belief-forming method is such that its reliability explains why we have it, but rather depending on whether we have established its reliability or not. Our first-order beliefs are justified, even if we have the second-order belief that the origin of the belief-forming method is such that its reliability does not explain why we have it, insofar as we have established its reliability.

This point applies not only to the example of the telescope but also to the example of the Bible and to the example of the psychological inclination. Your first-order religious beliefs are justified, even if you have the second-order belief that the origin of the Bible is such that its reliability does not explain why you rely on it, insofar as you have established its reliability. Similarly, the first-order beliefs that the psychological inclination produces are justified, even if we have the second-order belief that the origin of the psychological inclination is such that its reliability does not explain why we have it, insofar as we have established its reliability.

How can we establish the reliability of the psychological inclination? My answer to this question is simple and commonsensical. We can construct an enumerative induction for the psychological inclination, which holds that since some facts have explanations, most facts have explanations. Insofar as we have established the enumerative induction, we have established the reliability of the psychological inclination and the explainability principle.

To use an analogy, we present some white swans to justify the general belief that most swans are white. We do not argue that we have the second-order belief that the reliability of the psychological inclination to form the general belief would explain why we have the psychological inclination, and that the explanation indicates that we can trust the psychological inclination. By parity of reasoning, we should present some facts that have explanations to justify the explainability principle that most facts have explanations. We should not argue that we have the second-order belief that the reliability of the psychological inclination to form the explainability principle explains why we have it, and thus that we can trust it.

De Ray's debunking argument concludes that the explainability principle is not justified, which implies that we are not justified in believing that science has explanations. However, it is a plain fact that science has numerous explanations. For example, evolutionary theory has the explanation of why human embryos have gills, and the standard model has the explanation of why a quark has mass. Given that it is a plain fact that science has explanations, we are justified in believing that science has explanations. If a philosophical reflection upon evolutionary theory implies that we are not justified in believing that science has explanations, what is reduced to incredibility is not our belief that science has explanations but rather the philosophical reflection.

9.3.3 Self-Defeating

In this subsection, I argue that de Ray's debunking argument is self-defeating in a couple of respects.

As noted in Sect. 9.2 above, de Ray's debunking argument holds that we have the psychological inclination to believe that most facts have explanations because it was fitness-conducive, but not because it was reliable. Note that de Ray's debunking argument provides the evolutionary *explanation* of why we have the psychological inclination to establish that that we are not justified in believing that most facts have explanations. To use an analogy, de Ray provides a white swan to establish that we are not justified in believing that most swans are white. Such an argument is self-defeating.

There is another respect in which de Ray's debunking argument is self-defeating. It relies on evolutionary theory to conclude that the explainability principle is not justified. However, evolutionary theory explains biological phenomena, such as analogies and homologies. In my view, to appeal to evolutionary theory to conclude that the explainability principle is not justified is similar to providing some white swans to conclude that we are not justified in believing that most swans are white, which is self-defeating.

De Ray might retort that the explainability principle does not assert that all facts have explanations, but that most facts have explanations. No contradiction is involved in assuming that biological facts have explanations to argue that we are not justified in believing that most facts have explanations. Therefore, it is not self-defeating to rely on evolutionary theory to argue that the explainability principle is not justified.

This possible retort is ad hoc. It requires the specification of exactly what facts have explanations and what facts do not, and the justification for the distinction between these two groups of facts. It is not enough merely to attach the labels "non-brute facts" and "brute facts," respectively, to the former group of facts and the latter group of facts. After all, critics would switch the labels to the two groups of facts. It also requires the justification for classifying biological phenomena as belonging to the group of non-brute facts. In the absence of such justifications, it does not sound convincing to say that biological phenomena that evolutionary theory explains are non-brute facts while other phenomena are brute facts.

Let me turn to a related self-defeating claim that Darwin makes. He writes, "With me the horrid doubt always arises whether the convictions of man's mind, which has been developed from the mind of the lower animals, are of any value or at all trustworthy" (Darwin & Darwin, 1888). In other words, "the evolutionary origins of human minds ought to make us distrustful of such minds" (de Ray, 2022, p. 970). It is self-refuting for Darwin to say that human minds are not trustworthy due to their evolutionary origin. If he is right that human minds are not trustworthy, his mind is not trustworthy either, given that he is a human being too. As a result, we should distrust whatever he says, including evolutionary theory and the sentence just quoted above. It is ad hoc to say that all human minds, except Darwin's mind, are not trustworthy.

As Helen De Cruz et al. (2011, p. 518) note, it is self-defeating to rely on evolutionary theory, a product of our cognitive faculties, to establish skepticism about our cognitive faculties. Evolutionary theory presupposes that human minds are reliable. After all, evolutionary theory is a product of human minds. If human minds were unreliable, evolutionary theory would be not be justified. In consequence, it would be self-defeating for skeptics to appeal to evolutionary theory to establish skepticism about our cognitive faculties.

It is my assessment that Darwin and de Ray are similar to other thinkers: Thomas Kuhn (1962/1970, p. 172), van Fraassen (1980, pp. 39–40), Kathleen Akins (1996), and K. Brad Wray (2011, p. 8). These thinkers appeal to evolutionary theory to conclude, respectively, that science does not move toward truths, that the success of science can be explained from an antirealist point of view, that our cognitive faculties do not track truths, and that current scientific theories will be thrown out. Their philosophical conclusions are all undermined by their very appeal to evolutionary theory (Park, 2017).

9.4 Threat to Scientific Realism?

In this section, I argue that de Ray's debunking argument does not pose any threat to scientific realism, contrary to what he thinks.

De Ray thinks that his debunking argument sketched in Sect. 9.2 has a negative implication against scientific realism, which he defines as the view that "our best scientific theories accurately represent real, mind-independent states of affairs" (de Ray, 2022, p. 971). On the one hand, our best theories are the products of IBE, but the use of IBE requires the explainability principle. It follows that scientific realism is committed to the explainability principle (de Ray, 2022, pp. 971–972). On the other hand, scientific realism is committed to the truth of evolutionary theory which implies that the explainability principle is not justified. Therefore, "scientific realism undermines itself" (de Ray, 2022, p. 978).

In Sect. 9.3 above, however, I raised the following three objections to de Ray's debunking argument. (i) The two possibilities do not invalidate defenders' explanation. (ii) The explainability principle can be justified by the enumerative induction for it. (iii) It is self-defeating to appeal to evolutionary theory, which explains phenomena, to argue that we are not justified in believing that most facts have explanations. These three objections make it problematic for de Ray to say that it is self-defeating for scientific realism to be committed to the explainability principle and to the truth of evolutionary theory.

Let me turn to the no-miracles argument according to which the success of science would be inexplicable if our best theories were false, and thus they are true (Putnam, 1975, p. 73). Scientific realists might claim that the success of science justifies the reliability of science, and that the reliability of science explains why we trust science, i.e., that "the reliability of science as a source of beliefs about the world contributes to the explanation of why the realist relies on it" (de Ray, 2022, p. 983). Therefore, our best theories are justified.

De Ray raises the following objection to the no-miracles argument. The no-miracles argument sketched above instantiates IBE, but the use of IBE requires the explainability principle, and as noted in Sect. 9.2 above, the debunking argument concludes that the explainability principle is not justified. Therefore, the no-miracle argument "is unable to defend scientific realism from evolutionary scepticism" (de Ray, 2022, p. 984).

As noted in Sect. 4.4.3 of Chap. 4, however, it is to confuse a means and an end to say that the explainability principle is required for the use of IBE. To say so is similar to saying that to be on the roof is required for the use of a ladder. Just as to use the ladder is required to be on the roof rather than the other way around, so to use IBE is required for the explainability principle to be justified rather than the other way around.

9.5 The Two New Debunking Arguments

In this section, I sketch two debunking arguments that debunkers might construct in light of my objections to de Ray's debunking argument. In addition, I argue that they are better than de Ray's debunking argument and then raise objections to them.

The first new debunking argument concerns the six principles that we discussed in Sect. 5.3 of Chap. 5. They are the explainability principle, the uniformity principle that nature is uniform, the privilege principle that the best hypotheses are more likely to be chosen from good lots than from bad lots, the generalization principle that the world is generalizable, the simplicity principle that the world is simple, and the transparency principle that the appearance of theories tends to accord with their reality.

Why do we have the psychological tendency to believe the six principles? An evolutionary answer goes as follows. Our distant ancestors acquired the psychological tendency as a result of variations. It was fitness-conducive, i.e., those who had it did better in the battle to survive and reproduce than those who did not. It was passed on to next generations. Over many generations, those who had it persisted while those who did not perished. As a result, we have it. In short, we have it because it was fitness-conducive.

This new evolutionary explanation would be endorsed by debunkers and defenders alike. They would agree that the psychological tendency was fitness-conducive.

However, they would disagree on whether it tracked truths. Defenders would affirm, while debunkers would deny, that it was fitness-conducive because it tracked truths. Thus, debunkers' explanation holds that we have the psychological tendency because it was fitness-enhancing, while defenders' explanation holds that we have the psychological tendency because it was fitness-enhancing and it was fitness-enhancing because it tracked truths. Defenders' explanation indicates, while debunkers' explanation does not, that the six principles are justified.

How should we adjudicate between foregoing debunkers' explanation and defenders' explanation? Debunkers would argue that defenders' explanation assumes that the psychological tendency tracked truths while their explanation does not. Consequently, the simplicity principle dictates us to choose their explanation over defenders' explanation.[2] A corollary of choosing debunkers' explanation over defenders' explanation is that we are not justified in believing the six principles. Given that they are required for our best theories to be justified, we are not justified in believing our best theories.

Why is this new debunking argument better than de Ray's debunking argument? It is not vulnerable to my first objection to de Ray's debunking argument. My first objection was that the two possibilities do not invalidate defenders' explanation. In contrast to de Ray's debunking argument, the new debunking argument does not attempt to discredit defenders' explanation by merely saying that there might be no explanatory connection between defenders' explanans and the explanandum, and that the explanans might be false even if the explanandum is true, i.e., by merely saying that we might accidently have the psychological tendency to believe the six principles even if the psychological tendency was reliable, and that the psychological tendency could be fitness-conducive even if it was not reliable. Therefore, the new debunking argument is not subject to the objection that the two possibilities do not invalidate defenders' explanation that the psychological tendency was fitness-enhancing because it tracked truths.

The new debunking argument is not vulnerable to my second objection to de Ray's debunking argument either. My second objection was that the explainability principle can be justified by the enumerative induction for it. In contrast to de Ray's debunking argument, the new debunking argument does not claim that we have the second-order belief that the origin of the psychological tendency to believe the six principles is such that the reliability of the psychological tendency would not explain why we have the psychological tendency, and thus that the six principles are not justified. As a consequence, the new debunking argument is not subject to the objection that the six principles can be justified by relevant enumerative inductions for them.

By contrast, the new debunking argument is vulnerable to my third objection to de Ray's debunking argument. My third objection was that it is self-defeating to appeal to evolutionary theory, which explains phenomena, to argue that we are not

[2] Sharon Street (2006, p. 129) invokes the simplicity principle to choose debunkers' explanation for moral skepticism over defenders' explanation for moral realism.

justified in believing that most facts have explanations. Like de Ray's debunking argument, the new debunking argument appeals to evolutionary theory to argue that the six principles, including the explainability principle, are not justified, which is self-defeating.

The new debunking argument is self-defeating in another respect. It appeals to the simplicity principle to choose debunkers' explanation over defenders' explanation. However, it concludes that the six principles, including the simplicity principle, are not justified. It is self-defeating to appeal to the simplicity principle to argue that the simplicity principle is not justified.

The new debunking argument has another problem. Just as de Ray's debunking argument claims that the explainability principle is required for our best theories to be justified, so the new debunking argument claims that the six principles are required for our best theories to be justified. Accordingly, just as de Ray's debunking argument confounds a means and an end, so the new debunking argument confounds a means and an end.

These problems with the new debunking argument would motivate debunkers to construct another new debunking argument. It begins with the observation that we have the psychological propensity to explain and predict observables in terms of unobservables. It is due to this psychological propensity that we form scientific theories. Why do we have the psychological propensity? An evolutionary answer goes as follows. Our distant ancestors acquired it as a result of variations. The individuals who had it explained and predicted observables in terms of unobservables, while their competitors who did not have it did not. It was fitness-enhancing, and it was passed on to next generations. Over many generations, those who had it persisted while those who did not perished. As a result, we have it. In short, we have it because it was fitness-enhancing.

Again, this evolutionary explanation would be endorsed by debunkers and defenders alike. The only difference between them is that defenders would go further while debunkers would not, i.e., defenders would affirm, while debunkers would deny, that the psychological propensity was fitness-enhancing because it tracked truths. This evolutionary explanation indicates that our best theories are justified. The simplicity principle dictates us to choose debunkers' explanation over defenders' explanation. Therefore, we are not justified in believing our best theories.

The second new debunking argument is different from de Ray's debunking argument and the first new debunking argument in the following important respect. The second new debunking argument concludes directly that our best theories are not justified. By contrast, de Ray's debunking argument and the first new debunking argument conclude indirectly that our best theories are not justified. De Ray's debunking argument concludes that our best theories are not justified via concluding that the explainability principle is not justified, and the first new debunking argument concludes that our best theories are not justified via concluding that the six principles are not justified.

So what? The second new debunking argument is better than de Ray's debunking argument and the first new debunking argument. De Ray's debunking argument and the first new debunking argument attempt to show, respectively, that the explainability principle is not justified and to show that the six principles are not justified. These attempts are vulnerable to several objections unpacked in the previous sections and in this section. For example, the explainability principle can be justified by the enumerative induction for it, and it is self-defeating to appeal to the simplicity principle to argue that the simplicity principle is not justified. In contrast to de Ray's debunking argument and the first new debunking argument, the second new debunking argument makes no attempt to show that the six principles are not justified, and thus it is not vulnerable to the several objections.

In addition, de Ray's debunking argument and the first new debunking argument assert, respectively, that the explainability principle is required for our best theories to be justified, and that the six principles are required for our best theories to be justified. These assertions involve the confusion of means and ends. In contrast to de Ray's debunking argument and the first new debunking argument, the second new debunking argument is reticent about whether the six principles are required for our best theories, and thus it does not confuse the means and the end.

In this subsection, I sketched the two new debunking arguments not because I endorse them but rather because I believe that they are better than de Ray's debunking argument in multiple respects. I do not endorse even the second new debunking argument because like all the debunking arguments against scientific realism, it has two fundamental problems which are explicated in the next section.

9.6 Why Not Also Debunk the Simplicity Principle?

In this section, I expose a general problem with a debunking argument that appeals to the simplicity principle, whether it is directed at scientific realism or other philosophical views, such as moral realism and mathematical realism.

As noted in Sect. 9.5 above, the debunking argument against the simplicity principle is self-defeating, i.e., it is self-defeating to say that debunkers' explanation of why we have the psychological tendency to form the simplicity principle is simpler than defenders' explanation, and thus that the simplicity principle is not justified.

Suppose that a debunking argument relies on the simplicity principle, asserting that debunkers' explanation of why we have a cognitive faculty is simpler than defenders' explanation, and thus that a group of beliefs that the cognitive faculty produces is not justified. When debunkers advance such an argument, we can ask, "Why don't you also advance a debunking argument against the simplicity principle?"

This question creates the burden for debunkers to point out a relevant difference between the group of beliefs and the simplicity principle that would entitle them to advance the debunking argument against the group of beliefs, but not against the simplicity principle. Unless they specified the relevant difference, it would be arbitrary for them to advance the debunking argument against the group of beliefs, but not against the simplicity principle.

A debunking argument can be directed at the group of theoretical beliefs, the group of mathematical beliefs, the group of moral beliefs, and so on. Therefore, my objection that it is arbitrary to direct a debunking argument at a group of belief, but not at the simplicity principle, applies not only to a debunking argument against scientific realism but also to debunking arguments against moral realism, mathematical realism, and so forth.

9.7 A Reply to a Possible Move

In this section, I reply to a possible move that debunkers might make in light of my objection that it is self-defeating to appeal to evolutionary theory to discredit scientific realism.

Debunking arguments assume the truth of evolutionary theory. If evolutionary theory were false, debunking arguments, which are built upon the false theory, would be uncogent, and uncogent arguments can debunk nothing. However, it is self-defeating to assume the truth of evolutionary theory to discredit scientific realism, given that scientific realism claims that evolutionary theory is true. To appeal to the truth of evolutionary theory to discredit scientific realism amounts to appealing to the truth of evolutionary theory to discredit the truth of evolutionary theory.

Scientific antirealists, who believe that our best theories are empirically adequate, might argue that no theoretical term figures in evolutionary theory, thus that its truth coincides with its empirical adequacy, and that they can coherently make use of evolutionary theory to defend their philosophical view about science (Park, 2022, Sect. 5). Appealing to scientific antirealists' foregoing move, debunkers might claim that it is not self-defeating to appeal to the truth of evolutionary theory to argue against the truth of other scientific theories because to appeal to the truth of evolutionary theory is the same as appealing to the empirical adequacy of it.

Unfortunately, theoretical terms do figure in evolutionary theory. The evolutionary explanations of biological phenomena include many theoretical claims one of which is that "genotypic variation is generated by genetic mutations, i.e. changes to the DNA sequence leading to the emergence of new genotypes" (de Ray, 2022, p. 974). Such theoretical claims go beyond the empirical adequacy of evolutionary theory. Accordingly, the truth of evolutionary theory goes beyond the empirical adequacy of it.

9.8 Conclusion

My three objections to de Ray's debunking argument are as follows. The two possibilities do not invalidate defenders' explanation. The explainability principle can be justified by the enumerative induction for it. It is self-defeating to appeal to evolutionary theory, which explains phenomena, to discredit the explainability principle.

De Ray contends that scientific realism is self-defeating because it is committed to the explainability principle and to evolutionary theory which implies that the explainability principle is not justified. My response to this contention is to say that it is self-defeating for de Ray to appeal to evolutionary theory to discredit scientific realism because to appeal to evolutionary theory to discredit scientific realism amounts to appealing to evolutionary theory to discredit evolutionary theory.

My two new debunking arguments are better than de Ray's debunking argument. The first one holds that we have the psychological tendency to form the six principles because it was fitness-conducive. The second one holds that we have the psychological propensity to explain observables in terms of unobservables because it was fitness-conducive. The second one is better than the first one.

However, I endorse no debunking argument for the following reason. Any debunking argument relies on the simplicity principle, and we can construct a debunking argument whose conclusion is that the simplicity principle is not justified. It is arbitrary for debunkers to run a debunking argument against certain philosophical views, such as scientific realism, mathematical realism, and moral realism, but not against the simplicity principle.

This chapter has the following three key messages to all debunkers: "Mere possibilities debunk nothing," "Don't debunk evolutionary theory," and "Why not also debunk the simplicity principle?"

References

Akins, K. (1996). Of sensory systems and the 'aboutness' of mental states. *Journal of Philosophy, 93*(3), 337–372.

Darwin, C., & Darwin, F. (1888). *The life and letters of Charles Darwin*. John Murray.

De Cruz, H., Boudry, M., De Smedt, J., & Blancke, S. (2011). Evolutionary Approaches to Epistemic Justification. *Dialectica, 65*(4), 517–535.

de Ray, C. (2022). An evolutionary skeptical challenge to scientific realism. *Erkenntnis, 87*(2), 969–989.

Friedman, M. (1974). Explanation and scientific understanding. *The Journal of Philosophy, 71*, 5–19.

Goldman, A., & Beddor, B. (2021). Reliabilist epistemology. In E. N. Zalta (Ed.), *The Stanford encyclopedia of philosophy*. https://plato.stanford.edu/archives/sum2021/entries/reliabilism/

Hempel, C. (1966). *Philosophy of natural science*. Prentice-Hall.

Kuhn, T. (1962/1970). *The structure of scientific revolutions*. University of Chicago Press.

Laudan, L. (1977). *Progress and its problems: Towards a theory of scientific growth*. University of California Press.

Park, S. (2017). Problems with using evolutionary theory in philosophy. *Axiomathes, 27*(3), 321–332.

Park, S. (2022). The realist and selectionist explanations for the success of science. *Synthese, 200*(233). https://doi.org/10.1007/s11229-022-03703-0

Putnam, H. (1975). *Mathematics, matter and method: Philosophical papers* (Vol. 1). Cambridge University Press.

Salmon, W. (1978). Why ask, "Why?"? An inquiry concerning scientific explanation. In J. Kourany (Ed.), *Scientific knowledge: Basic issues in the philosophy of science*. Wadsworth Publishing Company.
Stanford, K. (2006). *Exceeding our grasp: Science, history, and the problem of unconceived alternatives*. Oxford University Press.
Street, S. (2006). A Darwinian dilemma for realist theories of value. *Philosophical Studies, 127*(1), 109–166.
van Fraassen, B. (1980). *The scientific image*. Oxford University Press.
van Fraassen, B. (1989). *Laws and symmetry*. Oxford University Press.
Wray, K. (2011). *Kuhn's evolutionary social epistemology*. Cambridge University Press.

Chapter 10
Merging the Moral Realism Debate with the Scientific Realism Debate

Abstract This chapter aims to enrich the moral realism debate with three classic arguments from the scientific realism debate. Just as scientific realists construct the no-miracles argument for scientific realism, so moral realists can construct the no-miracles argument for moral realism. Just as scientific antirealists construct the pessimistic induction and the problem of underdetermination for scientific antirealism, so moral antirealists can construct the pessimistic induction and the problem of underdetermination for moral antirealism.

Keywords Moral realism · No-miracles argument · Pessimistic induction · Scientific realism · Underdetermination

10.1 Introduction

This chapter discusses three classic arguments in the scientific realism debate, and then explores how moral realists and antirealists can make use of them to defend and attack moral realism, thereby providing readers with new perspectives for answering the following questions. Should scientific realists accept moral realism? Or can scientific realists reject moral realism? Is moral realism tenable? How about moral antirealism?

The plan for this chapter is as follows. In Sect. 10.2, I define scientific and moral realism. In Sect. 10.3, I construct the no-miracles argument (NMA) for moral realism that mirrors scientific realists' (Putnam, 1975, p. 73) NMA for scientific realism, and then sketch a possible debate between moral realists and antirealists regarding NMA for moral realism. In Sect. 10.4, I construct the pessimistic induction (PI) against moral realism that parallels scientific pessimists' (Laudan, 1977, p. 126) PI against scientific realism, and then sketch a possible debate between moral realists and antirealists regarding PI against moral realism. In Sect. 10.5, I claim that the problem of underdetermination occurs not only in science but also in morality, and then sketch a possible debate between moral realists and antirealists regarding the problem of underdetermination. In Sect. 10.6, I introduce an argument

S. Park, *Induction, Science, and Morality*, Synthese Library 520,
https://doi.org/10.1007/978-3-032-00032-3_10

for accepting scientific realism but rejecting moral realism. In each of these sections, I make various claims both from moral realists' perspective and moral antirealists' perspective. Those claims, some of which are bold and underdefended, are intended to serve as foils for future researchers to develop their own positions.

I have chosen NMA, PI, and the problem of underdetermination because they are classic arguments in the scientific realism debate. The space concern prevents me from discussing other arguments for and against scientific realism (Park, 2022a).

10.2 The Definitions of Scientific and Moral Realism

In this section, I define scientific and moral realism. These definitions will serve as the foundation for the discussion in the remainder of this chapter.

Scientific realism can be defined as the conjunction of the semantic thesis that scientific theories are true or false, the metaphysical thesis that some theoretical facts, which are mind-independent, inhabit the world, and the epistemological thesis that we know that some scientific theories are (approximately) true. In short, scientific realism holds that we know that some scientific theories correspond to the mind-independent parts of the world that we cannot observe with our naked eyes.

Similarly, moral realism can be defined as the conjunction of the semantic thesis that moral judgments are true or false, the metaphysical thesis that some moral facts, which are mind-independent, inhabit the world, and the epistemological thesis that we know that some moral statements correspond to moral facts. In short, moral realism holds that we know that some moral statements are made true by the mind-independent parts of the world.

We can distinguish between two versions of moral realism: the concrete and abstract versions. The concrete version holds that moral facts inhabit the concrete world in which there are spatiotemporal objects, such as mountains and black holes. The abstract version holds that moral facts inhabit the abstract world (Plato's heaven) in which there are aspatiotemporal objects, such as mathematical objects and propositions. This chapter concerns the concrete version, and Chap. 12 concerns the abstract version.

Scientific antirealism can be defined as any view that denies scientific realism. There are multiple variants of scientific antirealism in the literature, some of which claim that a theory is neither true nor false but rather useful or useless (Stanford, 2006, p. 189), that our best theories are false (Wray, 2013, p. 4327; Alai, 2017, p. 3282), that we do not know whether they are true or false, and that our best theories are closer to truths than their rivals, but not close to truths (Mizrahi, 2020).

Moral antirealism can be defined as any position that denies moral realism, or any position that does not assert that there are moral facts. There are multiple variants of moral antirealism in the literature. Emotivism (Stevenson, 1944), prescriptivism (Hare 1952), error theory Mackie (1977), quasi-realism (Blackburn, 1993), moral skepticism (Street, 2006) are different variants of moral antirealism. A variant of moral antirealism, which is not staked out by anyone, holds that current moral beliefs are closer to truths than their rivals, but not close to truths.

Scientific realists have varying views as to which theories in current science are true. Some suggest that successful ones are true (Putnam, 1975, p. 73). Some propose that theories making novel predictions are true (Leplin, 1997; Saatsi, 2009, p. 358). Others suggest that successful ones that fit together are true (Park, 2011a, 2019). In sum, different scientific realists take different theories in current science to be true.

Analogously, different moral realists would have different views on which moral statements are true. Hinduists and non-Hinduists might agree that some moral statements, such as the statement that charitable giving is (morally) right, are true. However, they might disagree on whether it is true or false that eating beef is right. I leave unresolved the issue of exactly which moral statements moral realists would take to be true.

Scientific and moral realism have a similarity: they share the view that our knowledge about the world can go beyond what we can observe. Scientific realism claims that we have knowledge about some theoretical facts. For example, we know that subatomic systems are the waves of probabilities. Like theoretical facts, moral facts are imperceptible, as David Hume (1978, pp. 468–469) claims. Moral realism claims that we have knowledge about moral facts although they are imperceptible. For example, we know that stealing has the property of wrongness although it is imperceptible.

Moral realists might argue that theoretical and moral facts are epistemically on a par, such that if we take the realist attitude toward theoretical facts, we should also take the realist attitude toward moral facts, and vice versa. It is arbitrary, they might maintain, to take the realist attitude toward theoretical facts but the antirealist attitude toward moral facts. This claim about the epistemic status of theoretical and moral facts echoes some mathematical realists' claim (Putnam, 1979, p. 347; Quine, 1980, p. 45) that it is arbitrary to take the realist attitude toward theoretical entities but the antirealist attitude toward mathematical entities. On these mathematical realists' account, just as empirical evidence justifies physical statements, so it justifies mathematical statements.

Scientific and moral antirealism have a similarity: they share the view that our knowledge about the world does not go beyond what we can observe. Scientific antirealism claims, for example, that we do not know that the nucleus of an atom consists of neutrons and protons, although we know that there are certain observable traces in a cloud chamber. Similarly, moral antirealism claims, for example, that we do not know that stealing has the property of wrongness, although we know that stealing occurs. Moral antirealists might argue that theoretical and moral facts are epistemically on a par, such that if we do not believe that theoretical facts exist, we should not believe either that moral facts exist, and vice versa.

How does moral realism relate to scientific antirealism? Moral realism asserts that we know about some moral facts although they are imperceptible. By contrast, scientific antirealism asserts that we do not know about theoretical facts because they are imperceptible. Consequently, to endorse moral realism and scientific antirealism at the same time incurs the burden of specifying a relevant difference between theoretical and moral facts that would justify the different epistemic attitudes toward them. What would be the reason for thinking that moral facts exist, but that theoretical facts do not?

This question can be the starting point of an interesting debate between rival discussants over whether theoretical and moral facts are real. K. Brad Wray (2013, p. 4327) and Mario Alai (2017, p. 3282) advocate scientific antirealism. If they advocated moral realism, they would have the burden of pointing out a relevant difference between theoretical and moral facts that would entitle them to take the antirealist attitude toward theoretical facts but the realist attitude toward moral facts. Carlo Alvaro (2020) advocates moral realism. If he advocated scientific antirealism, he would have the same burden as the one that Wray and Alai have.

How does moral antirealism relate to scientific realism? Moral antirealism claims that we do not know about moral facts because they are imperceptible, while scientific realism claims that we know about some theoretical facts although they are imperceptible. Accordingly, to embrace moral antirealism and scientific realism simultaneously incurs the burden of specifying a relevant difference between theoretical and moral facts or between scientific and moral hypotheses that would justify the different epistemic attitudes toward them. What would be the reason for thinking that we know about theoretical facts, but not about moral facts, or for thinking that we know that some scientific hypotheses are true, but not that some moral hypotheses are true? I answer this question in Sect. 10.6 below.

10.3 The No-Miracles Argument

In this section, I unpack the no-miracles argument (NMA) for scientific realism (Putnam, 1975, p. 73) and NMA for moral realism, and then sketch a possible debate between moral realists and antirealists regarding whether NMA for moral realism is tenable.

NMA for scientific realism goes as follows. Some theories are successful, i.e., they predict and explain many phenomena. For example, the Big Bang theory explains the cosmic background radiation and the Doppler red-shift of distant stars. It would take a miracle for false theories to be successful. By contrast, it would not take a miracle for true theories to be successful. A miracle is an inadequate explanans for the success of some theories. Therefore, scientific realism best explains why they are successful.

In a similar spirit, moral realists could run what I call "the no-miracles argument for moral realism." Some moral hypotheses are successful, i.e., they explain and predict many human actions. For example, the moral hypothesis that charitable giving is right explains why many people give to charity, and predicts that many people will give to charity. The moral hypothesis that stealing is wrong explains why many people refrain from stealing, and predicts that many people will refrain from stealing. How could such moral hypotheses be successful? What could explain their success? It would take a miracle for false moral hypotheses to be successful. By contrast, it would not take a miracle for true moral hypotheses to be successful. A miracle is an inadequate explanans for the success of some moral hypotheses. Therefore, moral realism, the view that successful moral hypotheses are true, best explains why they are successful.

This argument for moral realism can be alternatively formulated as follows. Many people act morally. It would take a miracle for them to act morally in the absence of some moral facts. By contrast, it would not take a miracle for them to act morally in the presence of some moral facts. To take an example, if the property of rightness did not inhere in the action of charitable giving, it would be a miracle that many people give to charity. If the property of wrongness did not inhere in stealing, it would be a miracle that many people refrain from stealing. A miracle is an unlikely event. Therefore, the existence of some moral facts best explains why many people act morally.

Now that we are clear about what NMA for moral realism says, we are ready to move on to a possible debate between moral realists and antirealists. Moral antirealists would appeal to scientific antirealists' objections to NMA for scientific realism to criticize NMA for moral realism, and moral realists would appeal to scientific realists' defenses of NMA for scientific realism to defend NMA for moral realism from moral antirealists' objections. The remainder of this section fleshes out this claim.

A standard objection to NMA for scientific realism is that there are counterexamples to it, i.e., that there were successful but false theories in the history of science. Scientific realism cannot explain why such theories were successful (Laudan, 1981). For example, the Ptolemaic theory, although false, explained and predicted the motions of celestial bodies. Aristotle's mechanics, although false, explained and predicted the motions of material objects. Such examples from the history of science are anomalies to the scientific realist explanation of why some theories are successful.

Moral antirealists would similarly object that there are counterexamples to the moral realist explanation of why some moral hypotheses are successful. For example, the Bible says that if a bride turns out to be a non-virgin, "the men of her town shall stone her to death" (Deuteronomy 22:21). The moral hypothesis that it is wrong for women to have sex before they get married was false, but it explained and predicted many human actions in the history of morality. So was the moral hypothesis that it is right to kill heretics. Such moral hypotheses from the history of morality are anomalies to the moral realist explanation that some moral hypotheses are successful because they are true.

Another standard objection to NMA for scientific realism is that it instantiates inference to the best explanation (IBE) (Laudan, 1981, p. 45; Fine, 1991, p. 82). Some scientific antirealists (van Fraassen, 1989) claim that IBE is not rationally compelling, i.e., that we can be rational even if we reject it. Such scientific antirealists would claim that we are rational even if we reject NMA for scientific realism. Appealing to this objection to NMA for scientific realism, moral antirealists might argue that NMA for moral realism instantiates IBE, and thus that it is not rationally compelling, and we are rational even if we reject it.

There are two responses available to moral realists. First, they might appeal to scientific realists' defense of IBE. In Sect. 4.4.3 of Chap. 4, for example, I argued that the argument from a bad lot, which is leveled at IBE, harbors the confusion of a means and an end. Moral realists could appeal to such defenses of IBE. Second, moral realists can admit that it is controversial whether IBE is rationally compelling, but argue that the two NMAs for scientific and moral realism are in the same boat, and thus that if NMA for scientific realism is correct, NMA for moral realism is also correct. This move is similar to some mathematical realists' (Baker, 2005, p. 225) response to scientific antirealists' objections to IBE. They argue that scientific and mathematical realism are in the same boat vis-à-vis the issue of whether IBE is compelling, and thus that if it is legitimate for scientific realists to use IBE, it is also legitimate for mathematical realists to use IBE.

Let me turn to another standard objection to NMA for scientific realism. Scientific antirealists have put forward several alternatives to NMA (Park, 2014). They all explain why some theories are successful, but none of them is committed to the truth of those theories. One of them holds, for example, that some theories make predictions because they are "empirically adequate" (Musgrave, 1988, p. 242). This explanation does not invoke the truth of theories, and hence it is useful to scientific antirealists, who aim to explain the success of theories without believing that they are true. The debate continues between scientific realists and antirealists over whether antirealist alternatives are tenable.

Analogously, moral antirealists could advance antirealist explanations of why some moral hypotheses are successful. In this vein, I propose a new metaethical theory that I call "cultural emotivism," according to which a moral statement is a description of a cultural emotion. For example, when moral evaluators, those who make moral judgments, say that charitable giving is right, they are expressing the belief that their culture has a positive emotion toward charitable giving. In addition, when they say that stealing is wrong, they are expressing the belief that their culture has a negative emotion toward stealing. When they describe the positive cultural emotion toward charitable giving, they might have a positive emotion toward charitable giving themselves; when they describe the negative cultural emotion toward stealing, they might have a negative emotion toward stealing themselves. However, their emotions are external to their descriptions of the cultural emotions. Finally, cultural emotivism affirms that there are cultural emotions toward certain actions, but denies that there are moral facts in the world. Accordingly, it is a form of moral antirealism.

There is a difference between cultural emotivism and Charles Stevenson's (1944) emotivism. Emotivism asserts that a moral statement is an expression of the speaker's emotion, while cultural emotivism asserts that a moral statement is the speaker's description of a cultural emotion. An emotion is neither true nor false, whereas a description of an emotion is either true or false. In consequence, emotivism denies, while cultural emotivism affirms, that a moral utterance is either true or false. On the cultural emotivist account, a moral utterance is true if and only if it corresponds to a cultural emotion, i.e., if and only if it captures the emotion of utterers' culture. To go further, we know that some moral utterances, such as the utterance that charitable giving is right, correspond to cultural emotions, i.e., that they are true.

There are also similarities and differences between cultural emotivism and cultural relativism (Park, 2011b). A similarity between them is that they share the view that a culture provides the standards for evaluation. On cultural relativism, an action is evaluated as being right or wrong, depending on whether it is accepted or rejected by a culture. Similarly, on cultural emotivism, a moral judgment is evaluated as being true or false, depending on whether it corresponds to a cultural emotion. A difference between cultural emotivism and cultural relativism is that cultural emotivism is a metaethical theory, a theory of what a moral judgment is, while cultural relativism is a normative ethical theory, a theory of what makes an action right or wrong. Cultural relativism claims that "cultural approval is what makes an act right, and cultural disapproval is what makes an act wrong" (Park, 2011b, p. 160).

How would cultural emotivists explain the success of some moral hypotheses? Consider the moral hypothesis that charitable giving is right. It is successful in that it makes the true prediction that many people will give to charity. How could it make the true prediction? On the cultural emotivist account, it makes the true prediction because it is true of a cultural emotion toward charitable giving, i.e., because it has captured a cultural emotion toward charitable giving. This explanation does not invoke a moral property that allegedly inheres in the action of charitable giving; it rather invokes a cultural emotion toward charitable giving. Therefore, it is a moral antirealist alternative to the moral realist explanation that the moral hypothesis that charitable giving is right makes the true prediction because it has captured a moral property that inheres in the action of charitable giving.

Just as cultural emotivism explains the success of some moral hypotheses, so other variants of moral antirealism—emotivism (Stevenson, 1944), prescriptivism (Hare, 1952), error theory (Mackie, 1977), quasi-realism (Blackburn, 1993), and moral skepticism (Street, 2006)—can also explain the success of some moral hypotheses. I do not unpack their explanations in the interest of saving space. Suffice it to say that there can be debates between moral realists and antirealists about which explanation is better.

10.4 The Pessimistic Induction

In this section, I argue that just as scientific pessimists construct PI against scientific realism, so moral pessimists can construct PI against moral realism, and that PI against moral realism is a prima facie threat to moral realism.

PI against scientific realism is the strongest argument for rejecting scientific realism (Mizrahi, 2020, p. 80). Its premise is that most old theories were rejected, and its conclusion is that most new theories will also be rejected. Pessimists provide a list of past theories to flesh out the premise. The most famous one is as follows:

Laudan's List

- the crystalline spheres of ancient and medieval astronomy;
- the humoral theory of medicine;
- the effluvial theory of static electricity;
- "catastrophist" geology, with its commitment to a universal (Noachian) deluge;
- the phlogiston theory of chemistry;
- the caloric theory of heat;
- the vibratory theory of heat;
- the vital force theories of physiology;
- the electromagnetic aether;
- the optical aether;
- the theory of circular inertia;
- theories of spontaneous generation. (Laudan, 1981, p. 33)

After putting forward this list, Larry Laudan claims that it "could be extended *ad nauseam*" (Laudan, 1981, p. 33). PI against scientific realism puts a damper on the optimistic outlook that we will obtain true scientific theories in the future.

Just as scientific pessimists construct PI against scientific realism, so moral pessimists can construct PI against moral realism. Its premise is that most moral statements that were widely accepted in the past were discarded, and its conclusion is that most moral statements that are widely accepted in the present will also be discarded.[1] The premise can be fleshed out by the following list:

The List of Past Moral Statements

- It is right to kill witches.
- It is right to commit genocide against heretics.
- It is right to enslave the conquered.
- It is right for slaves to obey their masters.
- It is right to stone a non-virgin bride to death.
- It is right to bind women's feet.
- It is right to bury servants alive along with a dead king.
- It is right to dismember rebels.
- It is right for widows to immolate themselves.
- It is right to kill prisoners of war on an altar to worship the sun.
- It is right for a man to have concubines.
- It is right to kill civilians during a war.
- It is right for a king to declare a war for his personal reasons.
- It is right to colonize another country.
- It is right to prohibit women from voting.
- It is right to commit honor killings.
- It is right for parents to choose their children's spouses.
- It is right to hunt animals for fun.

This list can be extended *ad nauseum* with more examples from the history of morality. PI against moral realism throws a wet blanket on the optimistic outlook that we will obtain true moral statements in the future.

[1] In this vein, Evans Williams writes, "most cultures have turned out to have major blind spots in their moral beliefs, and we are in much the same epistemic situation as they are, so we will probably also turn out to have major moral blind spots" (Williams, 2015, p. 974).

Imagine that our ancestors were moral realists, and that they believed that the property of rightness inhered in the action of enslaving the conquered. We can see in retrospect that they had a false belief about the action of enslaving the conquered. Imagine now that our ancestors were moral antirealists, and thus that they did not believe that the property of rightness inhered in the action of enslaving the conquered. We can see in retrospect that they had a true belief about the action of enslaving the conquered. The same story can be given mutatis mutandis about current moral realists and antirealists. Our descendants will see that current moral realists' beliefs about moral actions are false while current moral antirealists' beliefs about moral actions are true.

Moral realists would complain that we have no idea what moral statements will replace current moral statements, such as the statement that it is wrong for widows to immolate themselves. Moral pessimists would reply that the fact that we have no idea what moral statements will replace current moral statements only shows that we lack the cognitive capacity to conceive of alternatives to current moral statements. This reply echoes scientific pessimists' contention that the fact that we have no idea what scientific theories will replace current theories only shows that we lack the cognitive capacity to conceive of alternatives (Stanford, 2006, p. 45). Thus, just as current theories will be replaced by as-yet-unconceived alternatives, so current moral statements will be replaced by as-yet-unconceived alternatives.

In addition, moral pessimists would point out that homosexuality was permissible in ancient Greece. Ancient Greeks' judgment on homosexuality, although intuitive from our perspective, was later rejected. Women's social status was high during the Paleolithic Age. Our Paleolithic ancestors' judgment, although intuitive from our perspective, was later rejected. Such examples indicate that we cannot rule out the possibility that current moral judgments, which are intuitive from our perspective, might not be intuitive from our descendants' perspective and might later be rejected.

In response to these criticisms from moral pessimists, moral realists might appeal to scientific realists' responses to PI against scientific realism. Scientific realists have developed several strategies to tackle PI against scientific realism. Let me unpack some of them one by one how moral realists could utilize some of them, and how moral antirealists could respond to them.

Selective realists (Kitcher, 1993; Psillos, 1999) argue that some components of scientific theories succumbed to scientific revolutions but other components did not, and that the latter were used to predict and explain phenomena while the former were not. They call the former "idle posits" and the latter "working posits," and predict that working posits of currently accepted theories will similarly be retained in their future successors while idle posits of currently accepted theories will similarly be discarded.

In a similar spirit, moral realists might suggest that some components of past moral statements were retained in their current successors. However, this suggestion cannot get off the ground, given that a moral statement is not composed of multiple propositions but rather of a single proposition. For example, the past moral statement that it is right to kill witches is not composed of multiple propositions but rather of a single proposition. Therefore, a question does not even arise as to whether some of its constituent propositions survived a moral revolution.

Let me move on to another realist objection to PI against scientific realism. Some philosophers point out that Laudan's list above commits the fallacy of biased statistics. It is composed only of the past theories that were thrown out prior to the twentieth century. Most of the past theories that were widely accepted in the twentieth century have not yet been rejected in the twenty-first century. The number of scientific theories has increased dramatically in the twentieth century (Park, 2011c, p. 80; Mizrahi, 2013, 2015, 2016). Therefore, the premise of PI against scientific realism is false, and the conclusion of PI against scientific realism is not justified.

Moral realists might similarly suggest that most past moral statements have been stable, contrary to what the premise of PI against moral realism asserts. For example, it was considered to be right to give to charity, and it is still considered to be right to give to charity. The list of past moral statements introduced earlier in this section is biased in favor of discarded moral statements. A fair sample of past moral statements would include not only discarded past moral statements, such as the statement that it is right to kill witches, but also retained past moral statements, such as the statement that charitable giving is right.

In my view, this line of attack on PI against moral realism looks more promising than the selectivist line of attack on PI against moral realism sketched above. I leave to moral realists and antirealists the task of randomly selecting a large number of moral statements from the population of past moral statements. Such an empirical method would be useful to determine how plausible the premise of PI against moral realism is.

Let me turn to another realist response to PI against scientific realism. Counter-inductivists argue that "since past theories were false, present theories are true" (Park, 2018, p. 330). On this account, scientists learn from past mistakes concerning matters of fact and improve their theories. Failure is not the mother of more failures but rather the mother of success in science. Therefore, the fact that past theories were false is not evidence for the view that their current successors are false but rather evidence for the opposite view that their current successors are true.

Utilizing the counter-inductivist idea, moral realists could argue that since past moral statements were overturned, current moral statements will not be overturned. On this account, we learn from past mistakes concerning moral matters and improve our moral views. Failure is not the mother of more failures but rather the mother of success in morality. Therefore, the fact that past moral statements were false is not evidence for the view that current moral statements are false, but rather evidence for the opposite view that current moral statements are true.

In response, moral antirealists could say at best that PIs against scientific and moral realism are in the same boat, i.e., that if one is correct, the other is also correct, and that it is arbitrary to accept PI against scientific realism but to reject PI against moral realism. This response is analogous to some mathematical realists' contention that it is arbitrary to take the realist attitude toward theoretical entities but the antirealist attitude toward mathematical entities. It is also analogous to moral realists' possible contention sketched above that it is arbitrary to take the realist attitude toward some theoretical facts but the antirealist attitude toward some moral facts. I leave to future researchers the task of exploring whether there is any relevant difference between PI against scientific realism and PI against moral realism.

Let me turn to another realist response to PI against scientific realism. Some scientific realists (Musgrave, 1985, p. 211; Leplin, 1997, p. 141) argue that current theories are more successful than their predecessors that they displaced. For example, the general theory of relativity explains, while Newton's mechanics cannot, why gravitational waves are created. Therefore, the former is more likely to be true and/or closer to the truth than the latter, and it is fallacious to infer the demise of the former from that of the latter.

Can moral realists suggest that current moral statements are more successful than their predecessors? The correct answer to this question is elusive. Consider the old moral statement that it is right to commit genocide against heretics. This statement could explain and predict many phenomena in the past, but it can explain fewer phenomena in the present. Contrast it with its current successor that it is wrong to commit genocide against heretics. This statement can explain many phenomena in the present, but it could explain fewer phenomena in the past. It is not clear whether the new moral statemen is more successful than is the old moral statement.

10.5 The Problem of Underdetermination

The problem of underdetermination occurs not only in science but also in morality. The problem of underdetermination that occurs in science might be called "the problem of scientific underdetermination," and the problem of underdetermination that occurs in morality "the problem of moral underdetermination." In this section, I argue that just as the problem of scientific underdetermination favors scientific antirealism over realism, so the problem of moral underdetermination favors moral antirealism over realism.

The problem of scientific underdetermination occurs when observations cannot adjudicate between competing scientific theories. The rival theories make similar predictions about observables, but they make inconsistent claims about unobservables. For example, the Copenhagen and many-worlds interpretations of quantum mechanics make the same predictions about measurement outcomes. However, the former and the latter claim, respectively, that the wave function of a quantum system collapses and does not collapse upon measurement. Measurement outcomes cannot determine whether the wave function collapses.

Scientific realists and antirealists respond differently to the problem of scientific underdetermination (Park, 2009). Scientific realists choose the most virtuous theory and believe that it is likely to be true. A scientific theory is virtuous when it has virtues, such as simplicity, coherence, predictive power, explanatory power, and fruitfulness. By contrast, scientific antirealists choose the most convenient theory and believe that it is useful or empirically adequate. The problem of scientific underdetermination supports the choice of scientific antirealism over realism.

Let me turn to the problem of moral underdetermination. Gilbert Harman brings our attention to moral phenomena called "intractable moral disagreements" (Harman, 1996, p. 8). He notes that we can disagree about whether an action is right despite being provided with all the relevant observational data. For example, omnivorians and vegetarians believe, respectively, that it is morally permissible and not permissible to eat meat. Imagine that they together watch a person eat meat. Even if their faculties of perception function normally and flawlessly, they make opposing moral judgments about eating meat (Park, 2012, p. 190). Their moral disagreement is intractable.

In my view, the moral disagreement between vegetarians and omnivorians is similar to the factual disagreement between rival scientists espousing rival scientific theories. Just as rival scientists disagree about which of the rival scientific theories is true even if they are provided with all the relevant observational data, so vegetarians and omnivorians disagree about whether eating meat is right or wrong even if they are supplied with all the relevant observational data. The problem of underdetermination occurs both in science and in morality due to the fact that scientific theories and moral statements make claims about that which we cannot observe.

Vegetarians and omnivorians construct the following arguments for vegetarianism and omnivorism, respectively. Vegetarians argue that humans and animals are equal, given that they have the similar capacity to feel pain, and that just as it is wrong to kill humans for food, so it is wrong to kill animals for food (Singer, 2009). By contrast, omnivorians argue that humans and nonhuman omnivores, such as baboons and bears, are equal, and that just as nonhuman omnivores have the right to kill gazelles and buffalos for food, so humans have the right to kill livestock for food (Park, 2017, p. 292). Why do vegetarians and omnivorians construct such arguments?

Moral realists answer that the action of eating meat bears either the property of rightness or the property of wrongness, but not both. Vegetarians and omnivorians construct the aforementioned arguments for vegetarianism and omnivorism, respectively, to convince moral agents that the action of eating meat bears the property of wrongness and the property of rightness. The disagreement between them would be eliminated when they agree on which of the two properties the action of eating meat bears. So moral realists would depict the dispute between vegetarians and omnivorians.

Moral realists' foregoing description of the dispute between vegetarians and omnivorians is problematic. If moral realists are right that vegetarians and omnivorians attempt to show that the action of eating meat bears moral properties of wrongness and rightness, vegetarians and omnivorians should use scientific instruments, such as infrared detectors and electronic microscopes, to detect moral properties of wrongness and rightness, and then argue that their moral hypotheses have higher explanatory and predictive powers than their opponents' moral hypotheses. Specifically, vegetarians should argue that their moral hypothesis that eating meat is wrong explains and predicts more phenomena than omnivorians' moral hypothesis that eating meat is right; omnivorians should argue that their moral hypothesis that eating meat is right explains and predicts more phenomena than vegetarians' moral hypothesis that eating meat is wrong. It is mysterious why vegetarians and omnivorians construct, respectively, the argument that since all animals are equal, eating meat is wrong, and the argument that since all animals are equal, eating meat is right. What they do to justify their respective hypotheses is unintelligible under the moral realist framework, i.e., their actions of constructing the arguments are anomalies to moral realism. Moral realists have the burden of explaining why vegetarians and omnivorians construct the arguments instead of using scientific instruments and instead of arguing that their hypotheses have higher explanatory and predictive powers than their opponents' hypotheses.

Can moral antirealism explain why vegetarians and omnivorians construct the aforementioned arguments to justify their positions? There are several variants of moral antirealism in the metaethics literature. The notable ones are emotivism (Stevenson, 1944), prescriptivism (Hare, 1952), error theory (Mackie, 1977), quasi-realism (Blackburn, 1993), and moral skepticism (Street, 2006). These variants would provide different explanations of why vegetarians and omnivorians construct the aforementioned arguments to justify their positions. I unpack them in Sect. 15.4.6 of Chap. 15 in which I criticize error theory and moral skepticism. Suffice it to say here that moral realism and antirealism provide different explanations of why vegetarians and omnivorians construct arguments for vegetarianism and omnivorism, respectively.

10.6 Scientific Realism and Moral Antirealism

In this section, I introduce an argument for accepting scientific realism but rejecting moral realism.

How do scientists justify scientific hypotheses about theoretical entities, such as DNA and protons? They observe DNA with scientific instruments, such as electron microscopes. They perform experiments on protons with scientific equipment, such as particle accelerators. After doing such things, they argue that their hypotheses about theoretical entities are virtuous, e.g., their hypotheses have high explanatory and predictive powers.

Moral realists claim that some moral facts exist in the world, and that some moral statements are true in the way some scientific statements are true, i.e., that some moral statements are rendered true by moral facts in the world in the way some scientific statements are rendered true by scientific facts in the world. They claim, for example, that the action of torturing babies bears the property of wrongness, and that it is true that torturing babies is wrong because it corresponds to the moral fact that torturing babies is wrong.

Moral realists hypothesize that the action of torturing babies bears the property of wrongness. How do we know that their hypothesis true? How do we detect the property of wrongness that allegedly inheres in the action of torturing babies? Moral realists do not answer that we use scientific instruments, such as infrared detectors and electron microscopes. Nor do they answer that their moral hypothesis has high explanatory and predictive powers; they rather answer, to my surprise, that we use the faculty of reason (or ethical intuition) to know that their moral hypothesis is true, i.e., that the faculty of reason enables us to detect the property of wrongness in the action of torturing babies. The faculty of reason is allegedly a cognitive faculty that generates moral knowledge. Carlo Alvaro writes, for example, that "we can use reason and our moral experience of objective moral values and duties to determine that certain values are moral and immoral" (Alvaro, 2020, p. 32).

The foregoing account of how we detect a moral property in the world is vulnerable to the following two objections.

1. If it were legitimate for moral realists to hypothesize that we acquire moral knowledge with the use of the faculty of reason, it would also be legitimate for ghost-theorists to hypothesize that we acquire knowledge about ghosts with the use of the g-faculty. Just as it multiplies entities and faculties beyond necessity for ghost-theorists to postulate the g-faculty that can putatively detect ghosts in the world, so it multiplies properties and faculties beyond necessity for moral realists to postulate the faculty of reason that can putatively detect moral properties in the world. Just as ghost-theorists' account is merely music for those who already believe that ghosts exist, so moral realists' account is merely music for those who already believe that moral properties exist in the world (Park, 2022b, p. 272).
2. My hypothesis explaining why moral realists postulate the faculty of reason is that no moral property exists in the world; if the property of wrongness inhered in the action of torturing babies, moral realists would rather contend that we use scientific instruments to detect it in the action of torturing babies, and that their moral hypothesis that the action of torturing babies bears the property of wrongness has high explanatory and predictive powers. In short, moral antirealism explains why moral realists postulate the faculty of reason instead of using scientific instruments. Therefore, moral realists' method of defending moral realism "ironically indicates that there are no such things as moral knowledge and moral properties" (Park, 2022b, 275).

To use an analogy, imagine that you are a ghost-theorist. You aim to convince others that ghosts are real. Unfortunately, ghosts cannot be detected with the use of scientific instruments. Nor can your hypothesis that ghosts are real explain or predict any phenomena. Under such unfavorable circumstances, the natural thing for you to do is to postulate the g-faculty that putatively generates knowledge about ghosts. In my view, the hypothesis that ghosts are not real explains why ghost-theorists postulate the g-faculty. The same is true of moral realists. My hypothesis that no moral fact exists in the world explains why moral realists postulate the faculty of reason.

In case moral realists resist my foregoing case against moral realism, they have the burden of answering the following question. Why is it that there is a radical methodological difference between the way scientists know about scientific facts and their account of how we know about moral facts? Moral realists cannot answer that the radical methodological difference is due to the radical semantic difference between the way some scientific statements are rendered true and the way some moral statements are rendered true, for they earlier claimed that there is no such difference between them, i.e., they earlier claimed that just as some scientific statements are rendered true by scientific facts in the world, so some moral statements are rendered true by moral facts in the world.

How would moral realists respond to my forgoing objection to moral realism? They might reply that there are two versions of moral realism: the concrete and abstract versions. The concrete variant holds that we know that some moral statements correspond to moral facts in the concrete world, while the abstract variant holds that we know that some moral statements correspond to moral facts in the abstract world (Plato's heaven). On the concrete variant, a moral fact is a concrete fact, while on the abstract variant, a moral fact is an abstract one, just as a mathematical fact is an abstract one. Moreover, just as we use the faculty of reason to know about mathematical facts in the abstract world, so we use the faculty of reason to know about moral facts in the abstract world. I raise three objections to the abstract variant in Sect. 12.2.2 of Chap. 12.

Now, on what grounds would scientific realists resist moral realism? My brief answer to this question is that there is a wide gap between the way scientists justify scientific hypotheses and moral realists' account of how we justify moral hypotheses. Moral realists' account multiplies properties and faculties beyond necessity, and it can be reduced to absurdity, as the aforementioned story of how we detect ghosts illustrates. Therefore, it is not a double standard to accept scientific realism but to reject moral realism.

10.7 Conclusion

Just as scientific realism claims that some theories make true predictions because they are true of the world, so moral realism claims that some moral statements make true predictions because they are true of the world. However, cultural emotivism claims that some moral statements make true predictions because they are true of cultural emotions. There can be a debate between moral realists and cultural emotivists over which explanation is better.

Just as scientific antirealists construct PI and the problem of underdetermination against scientific realism, so moral antirealists can construct PI and the problem of underdetermination against moral realism. Just as scientific realists and antirealists debate over PI and the problem of underdetermination, so moral realists and antirealists can debate over PI and the problem of underdetermination.

Scientific realists do not need to accept moral realism because there is a radical difference between the way scientists justify scientific hypotheses and moral realists' account of how we justify moral hypotheses. Scientists use scientific instruments to confirm scientific hypotheses, and argue that scientific hypotheses have high explanatory and predictive powers. By contrast, moral realists hypothesize that we use the faculty of reason to detect some moral facts in the world. Their account multiplies properties and faculties beyond necessity, and can be explained by moral antirealism that no moral fact exists in the world.

It is a matter of future debates between moral realists and antirealists which ideas in this chapter are worth developing. In any event, I sum up this chapter in the following sentence: "The scientific realism debate can enrich the moral realism debate."

References

Alai, M. (2017). Resisting the historical objections to realism: Is Doppelt's a viable solution? *Synthese, 194*(9), 3267–3290.

Alvaro, C. (2020). The incoherence of moral relativism. *Cultura: International Journal of Philosophy of Culture and Axiology, 17*(1), 19–38.

Baker, A. (2005). Are there genuine mathematical explanations of physical phenomena? *Mind, 114*(454), 223–237.

Blackburn, S. (1993). *Essays in quasi-realism*. Oxford University Press.

Fine, A. (1991). Piecemeal realism. *Philosophical Studies, 61*(1–2), 79–96.

Hare, R. (1952). *Language of morals*. Clarendon Press.

Harman, G. (1996/2008). Moral relativism. In S. Satris (Ed.). *Taking sides: Clashing views on moral issues* (11th ed). Mcgraw-Hill/Dushkin.

Hume, D. (1978). In L. A. Selby-Bigge & P. H. Nidditch (Eds.), *A treatise of human nature*. Oxford University Press.

Kitcher, P. (1993). *The advancement of science: Science without legend, objectivity without illusions*. Oxford University Press.

Laudan, L. (1977). *Progress and its problems: Towards a theory of scientific growth*. University of California Press.

Laudan, L. (1981). A confutation of convergent realism. *Philosophy of Science, 48*(1), 19–49.

Leplin, J. (1997). *A novel defense of scientific realism*. Oxford University Press.

Mackie, J. (1977). *Ethics: Inventing right and wrong*. Penguin Books.

Mizrahi, M. (2013). The pessimistic induction: A bad argument gone too far. *Synthese, 190*(15), 3209–3226.

Mizrahi, M. (2015). Historical inductions: New cherries, same old cherry-picking. *International Studies in the Philosophy of Science, 29*(2), 129–148.

Mizrahi, M. (2016). The history of science as a graveyard of theories: A philosophers' myth. *International Studies in Philosophy of Science, 30*(3), 263–278.

Mizrahi, M. (2020). *The relativity of theory: Key positions and arguments in the contemporary scientific realism/antirealism debate*. Springer.

Musgrave, A. (1985). Realism vs. constructive empiricism. In P. Churchland & C. Hooker (Eds.), *Images of science: Essays on realism and empiricism* (pp. 197–221). University of Chicago Press.

Musgrave, A. (1988). The ultimate argument for scientific realism. In R. Nola (Ed.), *Relativism and realism in science* (pp. 229–252). Kluwer Academic Publishers.

Park, S. (2009). Philosophical responses to underdetermination in science. *Journal for General Philosophy of Science, 40*(1), 115–124.

Park, S. (2011a). Coherence of our best scientific theories. *Foundations of Science, 16*(1), 21–30.

Park, S. (2011b). Defence of cultural relativism. *Cultura. International Journal of Philosophy of Culture and Axiology, 8*(1), 159–170.

Park, S. (2011c). A confutation of the pessimistic induction. *Journal for General Philosophy of Science, 42*(1), 75–84.

Park, S. (2012). Against moral truths. *Cultura: International Journal of Philosophy of Culture and Axiology, 9*(1), 179–194.
Park, S. (2014). A pessimistic induction against scientific antirealism. *Organon F, 21*(1), 3–21.
Park, S. (2017). Moral vegetarianism vs. moral omnivorism. *Human Affairs, 27*(3), 289–300.
Park, S. (2018). The anti-induction for scientific realism. *Grazer Philosophische Studien, 95*(3), 329–342.
Park, S. (2019). The coherence of evolutionary theory with its neighboring theories. *Acta Biotheoretica, 67*(2), 87–102.
Park, S. (2022a). *Embracing scientific realism*. Springer.
Park, S. (2022b). Moral subjectivism vs. moral objectivism. *Filosofija Sociologija, 33*(3), 269–276.
Psillos, S. (1999). *Scientific realism: How science tracks truth*. Routledge.
Putnam, H. (1975). *Mathematics, matter and method: Philosophical papers* (Vol. 1). Cambridge University Press.
Putnam, H. (1979). *Mathematics, matter and method: Philosophical papers* (Vol. 1, 2nd ed.). Cambridge University Press.
Quine, W. (1980). *From a logical point of view* (2nd ed.). Harvard University Press.
Saatsi, J. (2009). Grasping at realist straws. *Metascience, 18*(3), 355–362.
Singer, P. (2009). *Animal liberation: The definitive classic of the animal movement*. HarperCollins Publisher.
Stanford, K. (2006). *Exceeding our grasp: Science, history, and the problem of unconceived alternatives*. Oxford University Press.
Stevenson, C. (1944). *Ethics and language*. Yale University Press.
Street, S. (2006). A Darwinian dilemma for realist theories of value. *Philosophical Studies, 127*(1), 109–166.
van Fraassen, B. (1989). *Laws and symmetry*. Oxford University Press.
Williams, E. (2015). The possibility of an ongoing moral catastrophe. *Ethical Theory and Moral Practice, 18*(5), 971–982.
Wray, B. (2013). Pessimistic induction and the exponential growth of science reassessed. *Synthese, 190*(18), 4321–4330.

Chapter 11
In Defense of Moral Functionalism

Abstract Moral functionalism holds that the essence of a moral judgment is its function of encouraging or discouraging moral actions. It is distinct from other metaethical theories, such as moral realism and emotivism. It is more effective than moral realism and error theory in preventing moral agents from committing morally wrong acts. In addition, it is immune to, while other metaethical theories are vulnerable to, the objection that we know better than do others about our own mental states.

Keywords Cognitivism · Exemplar · Function · Moral fact · Moral realism · Noncognitivism

11.1 Introduction

Can moral judgments be true or false? Do some moral facts exist? If they do, how do we know that they do? These are major questions in metaethics. Rival metaethicists provide rival answers to these questions. This chapter defends a new metaethical theory that I call "moral functionalism." It holds that the essence of a moral judgment lies in its function of encouraging or discouraging moral actions, actions that are either (morally) right or wrong. Moral functionalism provides answers to the above metaethical questions, and its answers are radically different from those provided by other metaethical theories.

My discussion of moral functionalism and other metaethical theories proceeds as follows. In Sect. 11.2, I expound upon moral functionalism, comparing it with moral realism. In Sect. 11.3, I compare moral functionalism with other metaethical theories: emotivism, prescriptivism, error theory, quasi-realism, moral skepticism, and cultural emotivism. It will become clear that moral functionalism is distinct from all of them. In Sect. 11.4, I construct two arguments for preferring moral functionalism to other metaethical theories. In Sect. 11.5, I reply to three possible objections to moral functionalism, one of which is that it is questionable whether moral judgments have an essential property.

S. Park, *Induction, Science, and Morality*, Synthese Library 520,
https://doi.org/10.1007/978-3-032-00032-3_11

Moral realism is the most popular metaethical view among academic philosophers. A survey of some academic philosophers finds that 56.4% of them endorse moral realism, 27.7% endorse moral antirealism, and 15.9% endorse other positions (Bourget & Chalmers, 2014, p. 476). Not surprisingly, moral realism will serve as the main foil for moral functionalism in this chapter.

This chapter is intended to be useful to those metaethicists who have qualms about existing metaethical theories, and to those metaethicists who are interested in how rival metaethicists can recast their metaethical theories in light of the objection that it is dubious that moral judgments have an essential property.

11.2 Moral Functionalism

In this section, I explicate moral functionalism by comparing it with moral realism.

In Sect. 10.2 of Chap. 10, I defined moral realism as the conjunction of the following semantic, metaphysical, and epistemological theses. The semantic thesis is that a moral utterance is true or false, i.e., that a moral judgment is a cognitive state. The metaphysical thesis is that some moral facts, which are mind-independent, exist. The epistemological thesis is that we know that some moral statements correspond to moral facts. Moral antirealism is any metaethical position that denies, or that does not claim, that some moral facts exist.

Imagine that intelligent aliens visit the earth to investigate moral phenomena. They observe that earthlings make the moral judgment that donating to charity is right, and the moral judgment that stealing is wrong. They set up rival hypotheses about the nature of moral judgments. The rival hypotheses make rival claims about whether moral judgments are cognitive states (beliefs) or noncognitive states (attitudes, emotions, and commands). However, they overlook that earthlings often admire and reward those who comply with the moral judgment that donating to charity is right, and that earthlings often despise and punish those who do not comply with the moral judgment that stealing is wrong. What are we to make of intelligent aliens' research on earthlings' moral judgments? In my view, it fails to capture the important aspect of moral judgments, viz., that moral judgments perform the functions of encouraging and discouraging moral actions, actions that are either (morally) right or wrong.

On the basis of this thought experiment, I define moral functionalism as the view that the essence of a moral judgment lies in its function of encouraging or discouraging moral actions. For example, the moral judgment that donating to charity is right essentially performs the function of encouraging moral agents, those who can commit right or wrong acts, to donate to charity. The moral judgment that stealing is wrong essentially performs the function of discouraging moral agents from stealing. In short, performing the function of influencing behaviors is the necessary and sufficient condition for a judgment to be moral.

On moral functionalism, moral utterances have indirect messages to moral agents that rewards or punishments might ensue. For example, when moral evaluators, those who make moral judgments, say, "Donating to charity is right," they are indirectly saying that moral agents might be admired and rewarded in case they comply with their moral judgment. When moral evaluators say, "Stealing is wrong," they are indirectly saying that moral agents might be despised and punished in case they do not comply with their moral judgment, i.e., they are indirectly issuing a warning to moral agents that stealing might incur contempt and punishment. In short, moral functionalism asserts that rewards and punishments might follow moral utterances.

On the moral functionalist account, a moral statement is similar to a legal statement in the following respect. Consider the legal statement "Stealing is illegal." It performs the function of discouraging potential perpetrators from stealing. When law enforcement officers say, "Stealing is illegal," they are indirectly saying that perpetrators might be convicted if they steal, i.e., they are indirectly issuing a warning to potential perpetrators that stealing might incur convictions. Similarly, consider the moral statement "Stealing is wrong." This moral statement performs the function of discouraging potential perpetrators from stealing. When moral evaluators say, "Stealing is wrong," they are indirectly saying that perpetrators might be despised and punished.

Moral functionalism is noncommittal regarding the semantic issue of whether a moral judgment is a cognitive state, regarding the metaphysical issue of whether some moral facts exist, and regarding the epistemological issue of whether we know that some moral statements are true. It does not even say that we do not know whether a moral judgment is a cognitive state, and whether some moral facts exist. Thus, moral functionalism is not a form of skepticism. However, it is a variant of moral antirealism, given that it fits the definition of moral antirealism provided earlier in this section: moral antirealism is any metaethical position that denies, or *does not claim*, that some moral facts exist.

An interesting consequence follows from the fact that moral functionalism is noncommittal regarding the semantic, metaphysical, and epistemological issues. Moral functionalism remains unscathed, even if cognitivists prove that moral judgments are cognitive states, even if noncognitivists prove that moral judgments are noncognitive states, even if moral realists prove that some moral facts exist, and even if moral antirealists prove that no moral fact exists. It is a misguided objection to moral functionalism to prove one way or the other that the semantic, metaphysical, and epistemological theses of moral realism are true or false.

Moral functionalism presupposes that the essence of a moral judgment is not what it is made up of but what it does. To use an analogy, the essence of money is not what it is made up of but what it does (Schick Jr. & Vaughn, 2010, p. 125). A bill and a coin are made up of different kinds of materials, but they are the same kind of thing, viz., money, because they basically perform the same functions, such as purchasing goods and repaying debts. In my view, the same holds for a moral judgment. A judgment is a moral one—whether it is a cognitive state or not—insofar as it performs the function of encouraging or discouraging moral actions.

It is not a new idea in philosophy to locate the essence of an object in its function rather than in its composition. For example, functionalism in philosophy of mind asserts that the essence of a mental state is what it does. Different kinds of physical things might perform the same function in human brains and alien brains, but as long as they perform the same function, they realize the same kind of mental state. By contrast, the identity theory asserts that the essence of a mental state is what it is made up of. Thus, even if two kinds of physical things perform the same function, they realize different kinds of mental states.

Let me apply moral functionalism to an example. Consider the following two moral sentences:

(D) Donating to charity is right.
(P) Donate to charity.

(D) is a descriptive sentence (or at least it has the form of a descriptive sentence), while (P) is a prescriptive sentence. Although they are of different sentential forms, they perform the same function of encouraging moral agents to donate to charity. It follows that they express the same moral judgment.

This view of moral sentences sits well with the fact that we can use either a descriptive sentence or a prescriptive sentence to induce the same action from others. Imagine, for example, that you and your spouse are in your apartment, that the window of your apartment is open, and that it is cold outside. Your spouse can speak either (D)' or (P)':

(D)' The window is open.
(P)' Close the window.

(D)' and (P)' perform the same function of encouraging you to close the window. Just as (D)' and (P)' perform the same function although they are of different sentential forms, so (D) and (P) above perform the same function although they are of different sentential forms.

11.3 Comparisons with Other Metaethical Theories

In this section, I compare moral functionalism with other metaethical theories to show that it is distinct from them.

As noted in Sect. 11.2 above, moral functionalism locates the essence of a moral judgment in what it does. It maintains that the essence of a moral judgment is what it does. By contrast, each of other metaethical theories to be examined in this section locates the essence of a moral judgment, as moral realism does, in what it is made up of. Let me go over them one by one.

Emotivism (Stevenson, 1944) asserts that moral statements are not descriptions of the world but expressions of emotions. They are neither true nor false, and no moral fact exists. For example, when we say that donating to charity is right, we are not ascribing the property of rightness to donating to charity but rather expressing a positive emotion toward donating to charity. A moral judgment is not a cognitive state—a mental state that represents the world—but rather a conative state—a mental state that motivates an action.

How does moral functionalism differ from emotivism? Emotivism claims that a moral utterance essentially expresses an emotion, which implies that an utterance that does not express an emotion is not a moral one. By contrast, moral functionalism claims that a moral utterance essentially performs the function of regulating moral actions, which implies that an utterance that does not perform the function is not a moral one. Moral functionalism does not require that a moral statement should be an expression of an emotion to perform the function of influencing behaviors. Nor does it require that it should be an expression of a belief to perform the function.

Let me turn to prescriptivism (Hare, 1952). It holds that a moral statement expresses a universalizable command, which can be applied to all moral agents in similar situations. For example, the moral statement "Stealing is wrong" is equivalent to the universalizable imperative statement "Don't steal." When you say, "Stealing is wrong," you are neither attributing the property of wrongness to stealing nor expressing an emotion toward stealing, but rather issuing the command that moral agents should not steal.

How does moral functionalism differ from prescriptivism? Prescriptivism asserts that a moral statement essentially expresses a command, which implies that a statement that does not express a command is not a moral one. By contrast, moral functionalism claims that a moral statement essentially performs the function of influencing behaviors. Moral functionalism does not require that a moral statement should be an expression of a command to perform the function of influencing behaviors.

Let me move on to error theory (Mackie, 1977). It asserts that a moral statement expresses a belief, but moral properties are queer in that they allegedly guide moral actions. Consequently, they are not real, and no moral belief is true. For example, the property of rightness does not inhere in the action of donating to charity, and hence it is false that donating to charity is right. The property of wrongness does not inhere in the action of stealing, and hence it is false that stealing is wrong. We are in error when we believe that some moral facts dwell in the world.

How does moral functionalism differ from error theory? Error theory is reticent about what functions moral judgments perform. By contrast, moral functionalism is reticent about whether moral utterances have truth-values, and whether some moral facts exist. As far as moral functionalists are concerned, moral judgments might or might not have truth-values, and moral properties might or might not exist.

Let me turn to quasi-realism (Blackburn, 1993). To put it roughly, quasi-realism holds that a moral judgement is a noncognitive state, that no moral fact exists, but that we can talk as if some moral judgments are true. We can say, for example, that it is true that stealing is wrong. To say so is equivalent to saying that stealing is wrong. Quasi-realism adopts the deflationary theory of truth, according to which saying that it is true that p is equivalent to saying that p.

How does quasi-realism differ from moral functionalism? Quasi-realism maintains that a moral judgment is a noncognitive state, and it invokes the deflationary theory of truth. In contrast, moral functionalism is neutral as to whether a moral judgment is a cognitive state, and it does not invoke any theory of truth. There is a similarity between them: both allow for the predicate "true" to be used in moral discourse. However, they do so on different grounds. Quasi-realism does so on the grounds that saying that it is true that p is equivalent to saying that p, while moral functionalism does so on the grounds that the moral statement that it is true that p basically performs the same function as the moral statement that p.

In my view, there is a rhetorical difference between the statement that it is true that p and the statement that p. In general, when you say that it is true that p, you are saying indirectly that your listeners should accept p whether they want to or not, because in general we should accept true propositions whether we want to or not (Park, 2012, p. 183). For example, when you say to witch-hunters that it is true that witches do not exist, you are indirectly saying that they should accept that witches do not exist whether they want to or not, because in general we should accept true propositions. Accordingly, when you say that it is true that stealing is wrong, you are indirectly saying that your listeners should accept that stealing is wrong whether they want to or not.

Moral skepticism (Street, 2006) holds that no moral judgment is justified. Some moral facts might or might not exist, but even if they do, we do not know that they do, which implies that we do not know whether some moral beliefs correspond to them. For example, there might or might not be a moral fact that donating to charity is right, but even if it does, we do not know that it does, which implies that we do not know whether the moral belief that donating to charity is right corresponds to it. In my view, moral skepticism is committed to cognitivism. To say that no moral judgment is justified presupposes that a moral judgment is a kind of mental state that can be shown to be likely to be true by evidence.

What are the differences between moral skepticism and moral functionalism? First, moral skepticism is committed to cognitivism, while moral functionalism is noncommittal about whether a moral judgment is a cognitive state. Second, moral skepticism is committed to the nonexistence of moral knowledge, while moral functionalism is noncommittal about the existence of moral knowledge. Third, moral skepticism is silent about what function a moral belief performs, while moral functionalism claims that it performs the function of encouraging or discouraging moral actions.

Cultural emotivism (Sect. 10.3 of Chap. 10) holds that the world is devoid of moral facts, and that a moral statement is a description of a cultural emotion. It is true or false not depending on whether it corresponds to a moral fact, but rather depending on whether it corresponds to a cultural emotion. For example, when you state that donating to charity is right, you are expressing the belief that your culture has a positive emotion toward donating to charity, and your belief is true or false depending on whether your culture has the positive emotion. When you state that stealing is wrong, you are expressing the belief that your culture has a negative emotion toward stealing, and your belief is true or false depending on whether your culture has the negative emotion.

How does moral functionalism differ from cultural emotivism? Cultural emotivism claims that a moral judgment is a cognitive state, while moral functionalism is neutral regarding whether it is a cognitive state, and hence whether a moral judgment corresponds to a cultural emotion. Cultural emotivism claims that a moral statement essentially describes a cultural emotion, which implies that a statement that does not describe a cultural emotion is not a moral one. By contrast, moral functionalism does not require that a moral statement should be a description of a cultural emotion to perform the function of regulating moral actions.

Moral functionalism competes with what I call "moral compositionalism," according to which the essence of a moral judgment lies in what it is composed of. All of the metaethical theories with which I have compared moral functionalism in this chapter are different forms of moral compositionalism, given that they all locate the essence of a moral judgment in what it is composed of. For example, moral realism and emotivism locate the essence of a moral judgment in a moral belief and a moral emotion, respectively. Thus, moral functionalism competes with all of the metaethical theories.

Moral functionalism maintains that just as it is wrong to require that money should be made up of a particular kind of material, so it is wrong to require that a moral judgment should be made up of a particular kind of mental state. In this regard, moral functionalism divorces itself from the diverse forms of compositionalism. All these diverse forms of compositionalism require that a moral judgment should be a particular kind of mental state. On their accounts, if a judgment is not made up of a particular kind of mental state, it is not a moral judgment.

11.4 Two Arguments for Moral Functionalism

11.4.1 Preventing Moral Agents from Committing Wrong Acts

In this subsection, I argue that moral functionalism is more effective than moral realism and error theory in preventing moral agents from committing wrong acts.

Imagine that religious fanatics resolve to kill heretics and openly assert that killing heretics is right, and that rival metaethicists resolve to prevent religious fanatics from killing heretics. Rival metaethicists would make the following rival utterances to religious fanatics:

Moral Realists: "Killing heretics is wrong. My moral belief is true, i.e., it corresponds to a moral fact."
Emotivists: "Killing heretics is wrong. I mean, killing heretics: Boo!"
Prescriptivists: "Killing heretics is wrong. I mean, don't kill heretics!"
Error Theorists: "Killing heretics is wrong. But my moral belief is false."
Quasi-Realists: "Killing heretics is wrong. I mean, killing heretics: Boo! It's true killing heretics is wrong."
Moral Skeptics: "Killing heretics is wrong. But my moral belief is not justified."
Cultural Emotivists: "Killing heretics is wrong. I'm describing the negative cultural emotion toward killing heretics."

Moral Functionalists: "Killing heretics is wrong. I'm discouraging you from killing heretics and indirectly saying you might be despised and punished in case you kill heretics."

These rival utterances have different degrees of effectiveness in preventing religious fanatics from killing heretics. How can we rank them in terms of the degrees of effectiveness? What would be their rankings?

Before I answer these questions, let me first state a caveat. Imagine that the president of the religious fanatics' country says to them, "If you kill heretics, you will be executed." Imagine also that molecular biologists have developed a gene-editing technique that predisposes humans to behave morally. No metaethicists' utterance can be as effective as the president's utterance or the gene-editing technique. Therefore, it is wrong to fault metaethicists' rival utterances above for not being as effective as the president's utterance or the gene-editing technique.

Error theorists' utterance is the least effective in preventing religious fanatics from killing heretics. To say to religious fanatics, "But my moral belief is false," does not help at all to dissuade them from killing heretics. Error theorists' utterance, if anything, adds fuel to religious fanatics' desire to kill heretics. Religious fanatics would say to error theorists, "You claim your moral belief is false. Why should I act on your false belief?"

Moral skeptics' utterance "But my moral belief is not justified" is slightly more effective than error theorists' utterance, given that moral skeptics leave the possibility open, while error theorists do not, that it might be true that killing heretics is wrong. Despite this advantage, however, moral skeptics' utterance is not effective at all in preventing religious fanatics from killing heretics. If anything, it adds fuel to religious fanatics' desire to kill heretics. Religious fanatics would say to moral skeptics, "You claim your moral belief is not justified. Why should I act on your unjustified belief?"

Moral realists' utterance "My moral belief is true" is more effective than moral skeptics' utterance. Moral realists' utterance implies that religious fanatics should accept that killing heretics is wrong whether they want to or not. However, moral realists' utterance is not effective in preventing religious fanatics from killing heretics. Religious fanatics would not care about whether their moral belief corresponds to a moral fact. All that moral realists can do under such circumstances is to wish that religious fanatics care about whether their moral belief corresponds to a moral fact.

Emotivists' utterance "I mean, killing heretics: Boo!" is more effective than moral realists' utterance. In general, an expression of anger toward an action has a more powerful impact on moral agents than does the expression of a belief about it, given that moral agents know that the expression of anger about an action is more frequently followed by punishment than is the expression of a belief about it. In other words, moral agents are more afraid of moral evaluators' anger than their belief.

Like emotivists' utterance, prescriptivists' utterance "I mean, don't kill heretics!" is more effective than moral realists' utterance. In general, the expression of a command has a more powerful impact on moral agents than does the expression of a belief, given that they know that the expression of a command is more frequently followed by punishment than is the expression of a belief. To put it differently, moral agents are more afraid of moral evaluators' command than their belief.

Quasi-realists' utterance "I mean, killing heretics: Boo! It's true killing heretics is wrong" is more effective than emotivists' utterance. Quasi-realists say, while emotivists do not, that it is true that killing heretics is wrong. The term "true" implies that religious fanatics should accept that killing heretics is wrong whether they want to or not. It is better for the purpose of preventing the killing of heretics to say that it is true that killing heretics is wrong than not to do so.

Cultural emotivists' utterance "I'm describing the negative cultural emotion toward killing heretics" is more effective than emotivists' utterance. When emotivists say, "Killing heretics is wrong," they are expressing their own negative emotion toward killing heretics. In contrast, when cultural emotivists say so, they are describing the negative cultural emotion toward killing heretics. Religious fanatics would be more afraid of the punishment from their culture than from emotivists.

Moral functionalists' utterance "I'm discouraging you from killing heretics and indirectly saying you might be despised and punished in case you kill heretics" is more effective than moral realists' utterance. Indirectly issuing the warning has a more powerful impact on religious fanatics' minds than saying that their moral belief does not correspond to a moral fact. Admittedly, religious fanatics might ignore the warning and do whatever they want to do. However, it is still true that they would care more about being despised and punished than holding a belief that does not correspond to a moral fact.

How would moral realists respond to my foregoing contention that moral functionalism is more effective than is moral realism in preventing religious fanatics from killing heretics? They might argue that when they say, "Killing heretics is wrong," they are also indirectly saying to religious fanatics that they might be despised and punished in case they kill heretics, i.e., they are also indirectly issuing the warning to religious fanatics, and accordingly moral realism is just as effective as moral functionalism in preventing moral agents from committing wrong acts.

In my view, however, it is extraneous to moral realism that a moral utterance indirectly issues a warning. Moral realism asserts that it is purely a cognitive affair to judge that killing heretics is wrong, just as it is purely a cognitive affair to believe that Pluto is round, and thus that to say that killing heretics is wrong is merely to attribute the property of wrongness to killing heretics, just as to say that Pluto is round is merely to attribute the property of roundness to Pluto. It is apropos of nothing that moral realists now claim that their moral utterance "Killing heretics is wrong" indirectly issues a warning.

By contrast, it is to be expected that moral functionalists say that their moral utterance "Killing heretics is wrong" indirectly issues a warning. Moral functionalism asserts that the moral utterance "Killing heretics is wrong" performs the function of discouraging religious fanatics from killing heretics. In general, we often despise and punish moral agents if they do what we have discouraged them from doing. Therefore, it is natural that moral functionalism asserts that a moral utterance indirectly issues a warning.

Moral realists would now object that the foregoing argument for the superiority of moral functionalism over moral realism is not epistemic but pragmatic. It shows that moral functionalism is more useful than is moral realism in preventing moral agents from committing wrong acts, but not that it is more likely to be true than is moral realism.

My response to this objection is to admit unabashedly that my foregoing argument for moral functionalism is a pragmatic one, but to call into question whether an epistemic argument for moral functionalism is needed. Those who doubt that a moral statement performs the function of influencing behaviors are asked to provide an account of why we utter moral statements, such as "Donating to charity is right" and "Stealing is wrong."

Moral realists might reply that we utter moral statements not because we want to regulate moral actions but because we feel good when we do it. This account of why we utter moral statements invites the following question. Why do we feel good when we express moral beliefs to others? My evolutionary answer is that our distant ancestors acquired the capacity to feel good when they expressed moral beliefs as a result of variations. They pursued the good feeling. As a result, they regulated each other's behaviors. By contrast, their competitors did not acquire this capacity. As a result, they did not regulate each other's behaviors. Over many generations, our ancestors persisted while their competitors perished. We inherited from our ancestors the gene which predisposes us to feel good when we express moral beliefs to others. This evolutionary explanation shows that we have been regulating each other's behaviors via pursuing the good feeling of expressing moral beliefs, and thus that expressing moral beliefs has been performing the function of influencing behaviors.

Let me turn back to the original objection that my argument for moral functionalism is a pragmatic one. It is not clear to me whether moral realists can construct a better pragmatic argument for moral realism. What is the use of moral realism? Why do moral realists aspire to establish moral realism? Moral realists would answer that they wish to secure the objectivity of morality, i.e., that they wish to show that some moral judgments are true or false not because we think that they are true or false but rather because they are made true or false by some moral facts. However, this answer invites a further question. What is the use of securing the objectivity of morality? As noted above, religious fanatics would not care about whether their moral belief corresponds to a moral fact.

Moral realists might now argue that moral realism is not meant to be useful but rather to be true. Specifically, it is not designed to encourage moral agents to do right things and to discourage them from doing wrong things, but rather to capture the world regarding the existence of some moral facts and to capture our minds regarding the existence of moral knowledge. To state it otherwise, it is not meant to influence moral agents' behaviors but to bring our attention to the existence of some moral facts in the world and to the existence of moral knowledge in our minds. Therefore, the fact that it is ineffective in preventing religious fanatics from killing heretics is not a strike against it.

My response to this possible move is to say that it does not matter whether moral realism is meant to be true or useful. The fact remains that moral realism is less useful than moral functionalism. Consider that the Copernican theory was not meant to explain the phases of Venus, although it was meant to explain the retrograde motions of planets. However, it unexpectedly explained the phases of Venus, while the Ptolemaic theory could not. As a result of such unexpected advantages, the Copernican theory defeated the Ptolemaic theory. It does not sound convincing to say that the Copernican theory was not meant to explain the phases of Venus, and that such unexpected advantages should not have contributed to the victory of the Copernican theory over the Ptolemaic theory.

There are moral phenomena that moral functionalism explains, but moral realism cannot. Moral evaluators sometimes punish moral agents. Imagine, for example, that religious fanatics killed heretics, and that moral evaluators punished religious fanatics. Why did moral evaluators punish religious fanatics? Moral functionalists answer that religious fanatics ignored the moral evaluators' warning, and that we usually punish others when they ignore our warnings. By contrast, moral realists do not have an answer to the question. From their point of view, it is mysterious why moral evaluators punished religious fanatics, for we do not usually punish others for holding beliefs that do not correspond to the world. Therefore, moral functionalism is better than moral realism, just as the Copernican theory was better than the Ptolemaic theory.

Michel Huemer (2016, pp. 2005–2006) claims that moral realism can explain, for example, why slavery was abolished. On his account, slavery was abolished ultimately because some of our modern ancestors recognized the moral fact that slavery was wrong. Those who endorse Huemer's defense of moral realism cannot say that moral realism is not meant to explain why moral evaluators sometimes punish moral agents. After all, it is arbitrary to say that moral realism is meant to explain why slavery was abolished, but not why moral evaluators sometimes punish moral agents.

11.4.2 The Privileged Access Objection

In this subsection, I argue that moral functionalism is not vulnerable to a standard objection to cognitivism and noncognitivism.

The standard objection relies on the idea (Goldman, 1993) that we know better about our mental states than do others, and conversely that others do not know better about our mental states than do we. If moral evaluators testify that their moral judgments are cognitive states, it is problematic for noncognitivists, as observers of moral evaluators, to maintain that moral evaluators' moral judgments are noncognitive states. After all, noncognitivists do not know better about moral evaluators' mental states than do moral evaluators themselves. In addition, if moral evaluators testify that their moral judgments are noncognitive states, it is problematic for cognitivists to maintain that moral evaluators' moral judgments are cognitive states.

After all, cognitivists do not know better about moral evaluators' moral judgments than do moral evaluators themselves. This objection to cognitivism and noncognitivism might be called "the privileged access objection."

The privileged access objection does not apply to moral functionalism, for moral functionalism is neutral as to whether a moral judgment is a cognitive state. Moral functionalism remains unrefuted whether moral evaluators testify that their moral judgments are cognitive or noncognitive states.

Objectors might argue that moral functionalism is subject to a similar objection, inviting us to imagine that moral evaluators contend that when they say, "Donating to charity is right," they do not *intend* their moral utterance to perform the function of encouraging the action of donating to charity. Moral functionalists, observers of moral evaluators, do not know better than do moral evaluators themselves about moral evaluators' intentions. Therefore, moral functionalism is flawed.

My first response to this objection is to point out that this thought experiment does not go hand in hand with the reality that we utter moral statements to affect behaviors. For example, when animal activists say, "Eating meat is wrong," they intend their moral statement to perform the function of discouraging moral agents from eating meat. It does not sound plausible to say that they do not intend their moral statement to perform the function of discouraging moral agents from eating meat. When moral evaluators make a moral utterance and claim that they do not intend it to perform the function of affecting behaviors, we can ask them the following question. Why utter the moral statement?

My second response to the objection is to point out that even if moral evaluators do not intend their moral utterance to perform a function, their moral utterance can nonetheless perform the function. Moreover, moral functionalists might know better than moral evaluators themselves about whether the moral statement performs the function. Consider, for example, that animal activists say, "Eating meat is wrong." Moral functionalists might closely watch, while animal activists do not, what animal activists' listeners do after they hear animal activists' utterance. As a result, moral functionalists might know, while animal activists do not, that listeners refrain from eating meat. Therefore, the privileged access objection does not apply to moral functionalism.

11.5 Objections and Replies

In this section, I reply to three possible objections to moral functionalism.

1. Other metaethicists might jointly ask the following question to moral functionalists. How can a theory of moral judgment be noncommittal about whether a moral judgment is a cognitive state? An informative theory of moral judgment should tell us one way or the other whether a moral judgment is a cognitive state.

   In my view, this question presupposes that the essence of a moral judgment is what it is made up of. Moral functionalists reject the presupposition and accept the alternative presupposition that the essence of a moral judgment is what it does. As far as moral functionalists are concerned, what a moral judgment does is more important than what it is made up of. To use the analogy of money above, what money does is more important than what it is made up of.

However, I do not claim that the debate between cognitivists and noncognitivists is worthless. The debate between them is valuable even if the essence of a moral judgment is its function of encouraging or discouraging moral actions. After all, the debate sheds light on what a moral judgment is made up of. As a result, it is useful to philosophers of mind who are interested in what a moral judgment is made up of.

The same holds for the debates over whether some moral facts exist and over whether we have moral knowledge. These debates cast light on what exists in the world and in our minds. As such, they are useful to metaphysicians who are interested in what exists in the world and to epistemologists who are interested in whether we have moral knowledge.

2. Critics might object that moral functionalism is undermined by the fact that some moral agents do not act on moral evaluators' moral judgments. For example, even if moral evaluators say to moral agents, "Donating to charity is right," most moral agents do not donate to charity. In addition, even if moral evaluators say to moral agents, "Stealing is wrong," some moral agents steal. Therefore, "Donating to charity is right" does not perform very well the function of encouraging moral agents to donate to charity, and "Stealing is wrong" does not very effectively discourage moral agents from stealing, contrary to what moral functionalism asserts.

   This objection to moral functionalism commits a straw man fallacy. Moral functionalism does not imply that the moral utterance "Donating to charity is right" is guaranteed to bring about the state of affairs in which all the listeners of the moral utterance donate to charity, or that the moral utterance "Stealing is wrong" is guaranteed to bring about the state of affairs in which none of the listeners of the moral utterance steals. In general, even if we encourage others to do something, they might not do what we encourage them to do; even if we discourage others from doing something, they might do what we discourage them from doing. It does not follow that we did not encourage them to do it or discourage them from doing it. It is wrong to fault moral functionalism for not being as effective as the gene-editing technique.

3. My opponents would provide counterexamples to moral functionalism, viz., moral judgments that do not perform the function of influencing behaviors, and nonmoral judgments that perform the function of influencing behaviors, thereby showing that the function of regulating behaviors is neither a necessary condition nor a sufficient condition for a judgment to be moral. My opponents would also provide some judgments at the borderline between moral and nonmoral judgments, and then challenge me to determine whether they are moral or nonmoral, thereby showing that it is questionable whether moral judgments have an essential property.

These objections to moral functionalism are in line with Stephen Stich's (2019) observation that many twentieth-century attempts to define a moral judgment in terms of necessary and sufficient conditions have failed. On his account, the concept of a moral judgment cannot be analyzed in terms of necessary and sufficient conditions, and we cannot draw a line between moral and nonmoral judgments. In consequence, he would reject moral functionalism on the grounds that it is built upon the dubious presupposition that the concept of a moral judgment can be analyzed in terms of necessary and sufficient conditions, and on the dubious presupposition that we can draw a line between moral and nonmoral judgments. In short, Stich would reject moral functionalism on the grounds that it is doubtful that moral judgments have an essential property.

My initial response to this possible objection is to admit that it is indeed a lethal objection to moral functionalism. I point out, though, that it applies not only to moral functionalism but also to other metaethical theories. For example, moral realism asserts that the essence of a moral judgment is the capability of being true or false of a moral action, which implies that if a judgment is not a belief about a moral action, it is not a moral judgment, and that if something is a belief about a moral action, it is a moral judgment. Moreover, belief comes in degrees. How strongly should we believe that an action is right to count as having a moral belief? The same holds for emotivism. Emotion comes in degrees. How strongly should we feel that an action is wrong to count as having a moral emotion? No answer to such questions is forthcoming. Therefore, all metaethicists should take Stich's possible objection seriously.

Should metaethicists despair and stop conducting research on moral judgments? My answer is "No." According to exemplar theory (Estes, 1986), a concept is not stored in our heads as necessary and sufficient conditions but as exemplars, and we do not classify objects with the use of necessary and sufficient conditions but with the use of exemplars. In my view, exemplar theory applies to moral judgments. We classify target judgments as moral or nonmoral, depending on whether they are sufficiently similar to an exemplar moral judgment.

Under the framework of exemplar theory, moral functionalists can reformulate their position as the view that a judgment is moral if and only if it is sufficiently similar to an exemplar moral judgment which performs the function of encouraging or discouraging moral actions. This version of moral functionalism might be called "the exemplar version." It contrasts with what I call "the essentialist version," according to which the essence of a moral judgment lies in its function of encouraging or discouraging moral actions. Thus, moral functionalism has essentialist and exemplar versions.

The same holds for other metaethical theories. For example, cultural emotivism has essentialist and exemplar variants. The essentialist variant holds that a moral judgment is essentially a belief about a cultural emotion. The exemplar variant holds that a judgment is moral if and only if it is sufficiently similar to an exemplar moral judgment which is a belief about a cultural emotion. The exemplar variants of metaethical theories go hand in hand with Stich's (1993) insight that we judge target moral actions to be right or wrong, depending on whether they are sufficiently similar to an exemplar right action or an exemplar wrong action.

After reformulating metaethical theories in terms of exemplar moral judgments, rival metaethicists can engage in the old debates about whether moral judgments are truth-apt, whether moral facts exist, and whether we can know that they do. In addition, they can engage in new debates about which of the newly formulated metaethical theories is the best, and about what moral judgments should be chosen as exemplar moral judgments. They can even attempt to develop new metaethical theories invoking exemplar moral judgments instead of necessary and sufficient conditions for a moral judgment. Therefore, Stich's possible objection to the essentialist version of moral functionalism has a positive impact rather than a negative impact on metaethics.

I anticipate that my critics will now challenge me to precisely define the predicate "sufficiently similar" that features in the exemplar version of moral functionalism. Similarity comes in degrees. Where should we draw a line between the judgments that are sufficiently similar to an exemplar moral judgment and the judgements that are not? Exactly how similar should target judgments be to an exemplar moral judgement to count as moral?

I make the following two critical comments regarding this challenge. First, it is built upon the dubious assumption that we can draw a line between objects to which a word correctly applies and objects to which a word incorrectly applies, and upon the dubious assumption that a concept can be defined in terms of necessary and sufficient conditions. Exemplar theory starts from the observation that no such line can be drawn, and from the observation that a concept cannot be analyzed in terms of necessary and sufficient conditions. Just as it is unfair to request that metaethicists should precisely define "moral judgment," so it is unfair to request that exemplar theorists should precisely define "sufficient similarity."

Second, an analysans for a vague analysandum does not need to be precise. In epistemology, for example, reliabilists analyze the concept of justification in terms of the reliability of cognitive process. They purposefully leave the analysans vague on the grounds that the analysandum is vague (Goldman & Beddor, 2021). To put it simply, they refuse to draw a line between reliable and unreliable cognitive processes on the grounds that we cannot draw a line between justified and unjustified beliefs. Similarly, I refuse to draw a line between judgments that are sufficiently similar to exemplar moral judgments and judgments that are not on the grounds that we cannot draw a line between moral and nonmoral judgments.

Consider also that vague predicates are useful under a certain condition. Bas van Fraassen says that "observables" and "unobservables" are usable predicates, although vague, insofar as we can provide clear examples of observables and unobservables (van Fraassen, 1980, p. 16). He says so in response to Grover Maxwell's (1962) objection that we cannot draw a line between observables and unobservables. Metaethicists can similarly say that "moral judgments" and "nonmoral judgments" are usable predicates, although vague, insofar as we can provide clear examples of moral and nonmoral judgments.

Moreover, vague predicates are useful in science. For example, medical science tells us that middle-aged women are liable to diseases, such as spinal stenosis and chronic bladder infection. Imagine that philosophers come along and say to medical scientists that we cannot draw a line between middle-aged women and non-middle-aged women, and that we cannot define "middle-aged women" in terms of necessary and sufficient conditions. Medical scientists would not think that they should stop conducting research on middle-aged women; they would rather think that they should stop talking with philosophers.

Vague predicates are also useful in philosophy. For example, epistemologists conduct research on justified beliefs, providing rival accounts of the conditions under which beliefs are justified. In a similar vein, metaethicists conduct research on moral judgments, developing theories about what moral judgments are composed of, what contributions they make to social changes (Sect. 13.5 of Chap. 13), and how they relate to the world. Metaethicists and philosophers in other fields of philosophy are doing useful things even if their key terms are vague.

11.6 Conclusion

Moral functionalism has essentialist and exemplar variants. The essentialist variant holds that a moral judgment essentially performs the function of encouraging or discouraging moral actions. The exemplar variant holds that target judgments are classified as moral or nonmoral depending on whether they are sufficiently similar to an exemplar moral judgment which performs the function of encouraging or the function of discouraging moral actions.

The common thread of the two versions is that what a moral judgment does is more important than what it is made up of, and that a moral judgment, be it a cognitive state or a noncognitive state, is a means of performing the function of regulating behaviors. As far as moral functionalists are concerned, cognitive states and non-cognitive states are different means of achieving the same end, viz., influencing behaviors. This chapter has the following simple message to the exponents of existing metaethical theories: "Observe what moral judgments do!"

References

Blackburn, S. (1993). *Essays in quasi-realism*. Oxford University Press.

Bourget, D., & Chalmers, D. (2014). What do philosophers believe? *Philosophical Studies, 170*(3), 465–500.

Estes, W. (1986). Array models for category learning. *Cognitive Psychology, 18*(4), 500–549.

Goldman, A. (1993). The psychology of folk psychology. *Behavioral and Brain Sciences, 16*(1), 15–28.

Goldman, A., & Beddor, B. (2021). Reliabilist epistemology. In E. N. Zalta (Ed.), *The Stanford encyclopedia of philosophy*. https://plato.stanford.edu/archives/sum2021/entries/reliabilism/

Hare, R. (1952). *Language of morals*. Clarendon Press.
Huemer, M. (2016). A liberal realist answer to debunking skeptics: the empirical case for realism. *Philosophical Studies, 173*(7), 1983–2010.
Mackie, J. (1977). *Ethics: Inventing right and wrong*. Penguin Books.
Maxwell, G. (1962). The ontological status of theoretical entities. In H. Feigl & G. Maxwell (Eds.), *Scientific explanation, space, and time: Minnesota studies in the philosophy of science* (pp. 181–192). University of Minnesota Press.
Park, S. (2012). Against moral truths. *Cultura: International Journal of Philosophy of Culture and Axiology, 9*(1), 179–194.
Schick, T., Jr., & Vaughn, L. (2010). *Doing philosophy: An introduction through thought experiments*. The McGraw-Hill Companies.
Stevenson, C. (1944). *Ethics and language*. Yale University Press.
Stich, S. (1993). Moral philosophy and mental representation. In M. Hechter, L. Nadel, & R. Michod (Eds.), *The origin of values* (pp. 215–228). Walter de Gruyter.
Stich, S. (2019). The quest for the boundaries of morality. In A. Zimmerman, K. Jones, & M. Timmons (Eds.), *The Routledge handbook of moral epistemology* (pp. 15–37). Routledge.
Street, S. (2006). A Darwinian dilemma for realist theories of value. *Philosophical Studies, 127*(1), 109–166.
van Fraassen, B. (1980). *The scientific image*. Oxford University Press.

Chapter 12
Does Mathematical Realism Help Moral Realism?

Abstract I offer three objections to mathematical and moral realism, according to which we know, respectively, that some mathematical and moral statements correspond to mathematical and moral facts in the abstract world. I also offer three objections to mathematical and moral fictionalism, according to which there are no mathematical and moral facts, respectively, and hence no mathematical and moral statements are true. I argue that mathematical inferentialism avoids the three objections to mathematical realism and the three objections to mathematical fictionalism, and that moral functionalism avoids the three objections to moral realism and the three objections to moral fictionalism.

Keywords Mathematical fictionalism · Mathematical inferentialism · Mathematical realism · Moral fictionalism · Moral functionalism · Moral realism

12.1 Introduction

In a survey of academic philosophers (Bourget & Chalmers, 2014, pp. 475–476), 39.3% of the respondents state that they accept Platonism, 37.7% nominalism, and 23.0% other positions. In addition, 56.4% of the respondents state that they accept moral realism, 27.7% moral antirealism, and 15.9% other positions. The survey indicates that mathematical and moral realism are popular among philosophers of mathematics and metaethicists, respectively.

The primary targets of this chapter are mathematical and moral realism, and the secondary targets are mathematical fictionalism (Balaguer, 2018) and moral fictionalism (Mackie, 1977), which are variants of mathematical and moral antirealism, respectively. The main thesis of this chapter is that just as mathematical inferentialism (Park, 2017, 2022a) has advantages over mathematical realism and fictionalism, so moral functionalism (Sects. 11.2 and 11.5 of Chap. 11) has advantages over moral realism and fictionalism. An implication of the main thesis is that moral realists have nothing to gain from appealing to mathematical realism.

S. Park, *Induction, Science, and Morality*, Synthese Library 520,
https://doi.org/10.1007/978-3-032-00032-3_12

The structure of the present chapter is as follows. In Sect. 12.2, I explicate mathematical and moral realism and then offer three objections to them. In Sect. 12.3, I explicate mathematical and moral fictionalism and then offer three objections to them. In Sect. 12.4, I argue that just as mathematical inferentialism avoids the three objections to mathematical realism and the three objections to mathematical fictionalism, so moral functionalism avoids the three objections to moral realism and the three objections to moral fictionalism. In Sect. 12.5, I reply to a possible objection to mathematical inferentialism.

The present chapter is intended to be beneficial to those who wonder whether moral realists can appeal to mathematical realism to defend moral realism, to those who wonder whether mathematical and moral fictionalism are tenable, and to those who wonder whether there is a new alternative to mathematical realism and fictionalism and whether there is a new alternative to moral realism and fictionalism.

12.2 Mathematical and Moral Realism

12.2.1 The Abstract World

In this subsection, I define mathematical and moral realism, which invoke the notion of the abstract world (Plato's heaven).

Mathematical realism holds that we know that some mathematical beliefs correspond to mathematical facts in the abstract world. Mathematical objects are nonspatiotemporal, noncausal, and immutable. We use the faculty of reason (or mathematical intuition) to apprehend mathematical facts in the abstract world. With the use of the faculty of reason, for example, we can know that the mathematical statement, $1 + 2 = 3$, corresponds to a mathematical fact in the abstract world.

Moral realism similarly suggests that we know that some moral beliefs correspond to moral facts in the abstract world. We use the faculty of reason (or ethical intuition) to grasp moral facts in the abstract world. Moral and mathematical facts are similar in that "Moral facts and mathematical facts are abstract entities, and as such, are different in kind from natural facts" (Kim, 2023). To put it another way, "moral values are abstract properties that are 'objective' in the same sense that geometrical or mathematical properties might be thought to be objective" (DeLapp, 2023).

This definition of moral realism differs from the definition of moral realism in Sect. 10.2 of Chap. 10, according to which we know that some moral statements correspond to moral facts in the concrete world. These two definitions of moral realism indicate that there are two versions of moral realism: the abstract version and the concrete version. The abstract version holds that we know that some moral beliefs correspond to moral facts in the abstract world, while the concrete version holds that we know that some moral beliefs correspond to moral facts in the concrete world. The concrete version was criticized in Sects. 10.2, 10.4, 10.5, and 10.6 of Chap. 10, and the abstract version is criticized in the present chapter.

Mathematical realists postulate mathematical facts to secure the objectivity of mathematics. The idea is that it is true that 1 + 2 = 3 not because we think that it is true but rather because there is a state of affairs in the abstract world that renders it true. If we believe that it is true, we have a true belief about that state of affairs. By contrast, if we believe that it is false, we have a false belief about that state of affairs. The truth-values of mathematical statements are just as objective as those of concrete statements, such as the statement that Jupiter has 82 moons.

Similarly, moral realists postulate moral facts to secure the objectivity of morality. The idea is that it is true that donating blood is (morally) right not because we think that it is true but rather because there is a state of affairs in the abstract world that renders it true. If we believe that it is true, we have a true belief about that state of affairs. By contrast, if we believe that it is false, we have a false belief about that state of affairs. The truth-values of moral statements are just as objective as those of mathematical statements.

The abstract version avoids David Hume's (1978, pp. 468–469) famous objection to the concrete version that moral properties are imperceptible. Imagine that you donate your blood in a hospital. The concrete version attributes the property of rightness to your action of donating blood. However, we cannot see the property of rightness in your action of donating blood. All we can see is you lying on a bed, a needle being pushed into your blood vessel, your blood flowing into a blood bag, and so on. The espousers of the abstract version would reply that we cannot see the moral property because it is abstract. We cannot see it any more than we can see mathematical objects, and yet it is just as real as mathematical objects.

12.2.2 Three Objections to Mathematical and Moral Realism

Mathematical and moral realism both postulate the abstract world and the faculty of reason, as noted in Sect. 12.2.1 above. In this subsection, I offer three objections that apply to both mathematical and moral realism.

What is the relationship between abstract and concrete objects? Plato's answer (Jones, 1970, p. 124) is that a concrete object participates in (or resembles) its corresponding abstract object. For example, a beautiful flower and a beautiful woman's face in the concrete world participate in the Form of Beauty in the abstract world. A concrete triangle drawn in ink on a piece of white paper participates in a triangle in the abstract world. Donating blood and saving a drowning child participate in the Form of Rightness in the abstract world.

Plato's foregoing view of the relationship between abstract and concrete objects is vulnerable to my following three objections.

(i) It is incomprehensible to say that the flower and the face resemble the Form of Beauty. After all, the Form of Beauty is not extended in space, which implies that it has neither a shape nor a size. It is not extended in time either, which implies that it has neither a beginning nor an end in time. We cannot visualize it, and we have no idea what it looks like. How can the flower and the face resemble such a thing? In short, it is unintelligible to say that spatiotemporal objects resemble nonspatiotemporal objects.

This objection also applies to the relationship between actions and the Form of Rightness. It is unintelligible to say that donating blood and saving a drowning child participate in the Form of Rightness. After all, the Form of Rightness is not spread out in space, which implies that it has neither a shape nor a size. It is not spread out in time either, which implies that it has neither a beginning nor an end in time. We cannot visualize it, and we have no idea what it looks like. How can donating blood and saving a drowning child participate in such a thing? In short, it is an inscrutable notion that concreta participate in abstracta.

We might grant for the sake of argument that we can make sense of the notion that concreta can participate in (or resemble) abstracta. Plato claims that a concrete object is what it is because it participates in an abstract object. For example, the flower and the face are beautiful because they participate in the Form of Beauty. What are we to make of this explanation of why the flower and the face are beautiful? My answer is that it puts the cart before the horse. This answer is fleshed out as follows.

In general, we should explain a relational property in terms of an intrinsic property rather than the other way around because a relational property depends on an intrinsic property for its existence (Park, 2003, p. 166). Consider, for example, two gold bars. One gold bar has the intrinsic property of being square and the relational property of being similar to the other gold bar with respect to shape. One gold bar is square even if the other gold bar does not exist, but one gold bar cannot be similar to the other gold bar with respect to shape without being square. As a consequence, it is accurate to say that one gold bar is similar to the other gold bar with respect to shape because it is square, but inaccurate to say that one gold bar is square because it is similar to the other gold bar with respect to shape.

The same holds for the relationship between the face and the Form of Beauty. The face has the intrinsic property of being beautiful and the relational property of being similar to the Form of Beauty. The face is beautiful even if the Form of Beauty does not exist, but it cannot be similar to the Form of Beauty without being beautiful. Accordingly, it is accurate to say that the face resembles the Form of Beauty because it is beautiful, but inaccurate to say that the face is beautiful because it resembles the Form of Beauty.

Initially, it sounds profound to say that the face is beautiful because it resembles the Form of Beauty. In my view, however, the source of this seeming profundity is not a metaphysical truth but the confusion of the explanatory relationship between intrinsic and relational properties. An intrinsic property and a relational property should play the roles of an explanans and an explanandum, respectively, but not the other way around. To state it otherwise, an intrinsic property should serve as an explanans for a relational property, but not the other way around.

This philosophical lesson also applies to the explanatory relationship between a concrete triangle drawn in ink on a piece of white paper and a triangle in the abstract world. The concrete triangle has the intrinsic property of being three-sided and the relational property of resembling the abstract triangle. The concrete triangle is three-sided even if the abstract triangle does not exist, but it cannot resemble the abstract triangle without being three-sided. Consequently, it is accurate to say that the concrete triangle resembles the abstract triangle because it is three-sided, but inaccurate to say that the concrete triangle is three-sided because it resembles the abstract triangle.

The same holds for the explanatory relationship between donating blood and the Form of Rightness. Donating blood has the intrinsic property of rightness and the relational property of participating in the Form of Rightness. Donating blood is right even if the Form of Rightness does not exist, but it cannot participate in the Form of Rightness without being right. Therefore, it is accurate to say that donating blood participates in the Form of Rightness because it is right, but inaccurate to say that donating blood is right because it participates in the Form of Rightness.

Mathematical and moral realists might suggest that a concrete object is what it is because it instantiates or exemplifies an abstract object. For example, a flower is beautiful because it instantiates the Form of Beauty. Unfortunately, this suggestion also puts the cart before the horse. It is accurate to say that a flower instantiates the Form of Beauty because it is beautiful, but inaccurate to say that the flower is beautiful because it instantiates the Form of Beauty. Imagine that someone asks why a flower instantiates the Form of Beauty, but not the Form of Ugliness. A natural response to this question is to say that a flower is beautiful, but not ugly.

(ii) It goes against Ockham's Razor for mathematical and moral realists to say that we have the faculty of reason that generates knowledge about abstract facts. If it were legitimate for mathematical and moral realists to say so, it would also be legitimate for unicorn-believers to say that we have the u-faculty that generates knowledge about unicorns, for ghost-believers to say that we have the g-faculty that generates knowledge about ghosts, and so on (Park, 2022a, p. 183, 2022b, p. 272). Just as unicorn-believers and ghost-believers multiply entities and faculties beyond necessity, so mathematical and moral realists multiply facts and faculties beyond necessity. It is problematic to multiply facts beyond necessity. It is all the more so to multiply facts and faculties beyond necessity.

To go further, I contend that we have the s-faculty (the super-faculty), whose sole function is to generate the knowledge that the abstract world does not exist, and hence that there are no abstracta, including mathematical and moral facts. The s-faculty is higher and more powerful than the faculty of reason. Accordingly, it can overturn the beliefs that the faculty of reason generates in our minds about the abstract world. Furthermore, it is infallible while the faculty of reason is fallible. Consequently, it never produces a false belief.

Opponents would object that mathematical and moral realists must have the s-faculty too, but they claim that the abstract world exists. Why do they believe that the abstract world exists? What happened to their s-faculty? My answer is that unlike other cognitive agents, mathematical and moral realists were born without the s-faculty (alternatively, they were born with the s-faculty, but their s-faculty is inactive). As they interact more and more with antirealists, their s-faculty will become active. This answer is merely an extension of moral realists' contention that some individuals' faculty of reason is more powerful than other individuals' faculty of reason, and that as a result, "some are better at apprehending moral truths than others" (Huemer, 2016, p. 2004).

Objectors would say that we have no independent reason for postulating the s-faculty. The s-faculty is postulated solely for the sake of undermining the realist account of how we acquire knowledge about the abstract world. However, this objection to my hypothesis of the s-faculty can be turned against mathematical and moral realists' hypothesis of the faculty of reason (or mathematical and ethical intuitions). We have no independent reason for positing the faculty of reason that putatively generates knowledge about abstract facts. The faculty of reason is posited solely for the sake of giving an epistemological account of how we grasp abstract facts.

Moreover, why is it that the faculty of reason produces knowledge about mathematical and moral facts in the abstract world, but not knowledge about concrete facts in the concrete world? Why is it, for example, that the faculty of reason produces the mathematical knowledge that there are infinitely many natural numbers in the abstract world, and the moral knowledge that the Form of Rightness is nonspatiotemporal, but not the concrete knowledge about exactly how many dragonflies and mosquitos there are in the concrete world? Mathematical and moral realists have the burden of answering this question (Park, 2022b, p. 272).

Under what conditions can we conclude that we have the faculty of reason that generates mathematical and moral knowledge about the abstract world? We could do so provided that we can calibrate it with the use of other cognitive faculties. To use an analogy, we can conclude that visual perception generates knowledge about physical objects because we can calibrate it with the use of auditory perception and tactile perception. In contrast to visual perception, we cannot calibrate the faculty of reason with the use of other cognitive faculties because we do not have other cognitive faculties that can be used to apprehend abstract facts. Accordingly, the epistemological story that we have the faculty of reason that generates mathematical and moral knowledge about abstract facts is merely music for those who already believe that we have knowledge about abstract facts (Park, 2022b, p. 272).

Let me now turn to what contemporary mathematical realists think of the abstract world and the faculty of reason. Like Plato, they think that the abstract world exists, but unlike Plato, they do not think that we have the faculty of reason that purportedly grasps abstract objects. Consider two famous contemporary arguments for mathematical realism: the Quine-Putnam indispensability argument (Quine, 1948; Putnam, 1971) and the enhanced indispensability argument (Baker, 2005, p. 225, 2009, p. 614). They both assert that mathematical entities are abstracta, that mathematical statements are descriptions of abstracta, but that some mathematical statements are confirmed by observations of the concrete world. On the indispensabilist account, experience is the source of justification for some mathematical statements, and it is intellectually dishonest to believe that experience justifies theoretical statements, but not mathematical statements, i.e., it is unfair to embrace scientific realism but to resist mathematical realism.

In the indispensabilist spirit, moral realists might suggest that moral statements are descriptions of the abstract world, but that some of them are confirmed by observations of the concrete world. What would be the reason for thinking that experience is the source of justification for some moral statements? Some moral statements can explain and predict human behaviors (Sect. 10.3 of Chap. 10). For example, the moral hypothesis that donating blood is right explains why many people donate blood and predicts that many people will donate blood. The moral hypothesis that stealing is wrong explains why many people refrain from stealing and predicts that many people will refrain from stealing. Since some moral hypotheses explain and predict human actions, moral realists might say that they are justified by experience.

What are we to make of the foregoing suggestions by indispensabilists and moral realists? They do not invoke the faculty of reason, and hence they avoid my earlier objection to mathematical and moral realism that to postulate the faculty of reason is to multiply faculties beyond necessity. However, they invoke the abstract world, and hence they are vulnerable to my objections to Plato's view unfolded earlier in this section. Specifically, it is an inscrutable claim that a concrete triangle drawn in ink on a piece of white paper can resemble a triangle in the abstract world. In addition, it is conceptually problematic to say that a concrete triangle is three-sided because it participates in an abstract triangle.

(iii) Indispensabilists face the following additional objection. If mathematical statements are descriptions of the abstract world, they do not make any assertions about the concrete world, and hence they cannot be confirmed by observations of the concrete world (Park, 2022a, p. 189). To use an analogy, imagine that astronomers claim that astronomical statements about Jupiter, e.g., that Jupiter has 82 moons, are confirmed by observations of human faces on the earth. You would be puzzled by these astronomers' claim. You should be similarly puzzled by indispensabilists' claim that mathematical statements, which are descriptions of the abstract world, are confirmed by observations of the concrete world.

A similar objection applies to moral realists. If moral statements are descriptions of the abstract world, they do not make any assertions about the concrete world, and hence they cannot be confirmed by observations of the concrete world. Observations of the concrete world cannot justify moral statements about the abstract world any more than observations of human faces on the earth can justify astronomical statements about Jupiter, or any more than observations of the concrete world can justify mathematical statements about the abstract world.

Mathematical and moral realists might reply that concreta participate in (or resemble) abstracta, and thus that we can know about abstracta by observing concreta. To use an analogy, we can know about our faces by observing their images in a mirror. Unfortunately, this reply is not convincing. As noted earlier in this section, it is mysterious how spatiotemporal objects can participate in (or resemble) nonspatiotemporal objects. Insofar as the relationship between concreta and abstracta remains elusive, it also remains elusive how we can know about abstracta by observing concreta.

Attaching the labels "participate in" and "resemble" to the relationship between concreta and abstracta does not help to illuminate the relationship between them. It is intelligible to say that a philosopher participates in a philosophy conference, but not that concreta participate in abstracta. After all, a philosopher takes up space, and a conference occurs in a certain region of space. We can visualize the state of affairs in which the philosopher enters a certain region of space where the conference is held. In contrast, an abstract triangle does not take up space. As a result, we cannot visualize the state of affairs in which a concrete triangle enters the abstract triangle. To take another example, it is intelligible to say that a coin is similar to another coin, given that they both occupy space and they both are round. By contrast, donating blood takes up space, but the Form of Rightness does not. As a result, we have no idea how donating blood can be similar to the Form of Rightness.

Mathematical and moral realists might suggest that the relationship between abstracta and concreta is causal, i.e., that without abstracta, concreta could not exist, and thus that it is problematic for me to say that saving a drowning child can be right even if the Form of Rightness does not exist. Unfortunately, this suggestion is problematic. Recall that abstracta are nonspatiotemporal, noncausal, and immutable, while concreta are spatiotemporal, causal, and mutable. On these definitions of abstracta and concreta, there can be no causal relationship between abstracta and concreta, and it is problematic to suggest that without the Form of rightness, no action would be right.

In this subsection, I unpacked three objections to both mathematical and moral realism. They present challenges to those who argue that mathematical and moral realism are in the same boat, that mathematical realism is tenable, and thus that moral realism is also tenable.

12.2.3 A Possible Move

In the previous subsections of this section, I offered three objections that apply to both mathematical and moral realism. In this subsection, I reply to a possible move that moral realists might make in the face of those objections to moral realism.

Moral realists might give up the abstract variant of moral realism and instead embrace the concrete variant of moral realism. An important difference between the two variants is that the abstract variant holds that moral facts are abstracta while the concrete variant holds that moral facts are concreta. On the concrete variant, moral facts exist in the concrete world, and actions bear properties of rightness and wrongness. In addition, moral statements are rendered true by moral facts in the way concrete statements, such as the statement that Jupiter has 82 moons, are rendered true by concrete facts in the concrete world.

What are we to make of the concrete variant? I have offered several objections to it in Sects. 10.2, 10.4, 10.5, and 10.6 of Chap. 10. In this subsection, I offer a new objection to it, which is inspired by Hume's (1978, pp. 468–469) view that a moral property is imperceptible. Imagine that you donate your blood in a hospital. You are lying on a bed. A needle is pushed into your vessel. Your blood flows from your vessel into a blood bag. According to the concrete variant, your action of donating blood bears the property of rightness. However, it is not clear exactly where the moral property exists. Does it exist in your whole body, in your brain, in your blood vessel, or in the blood bag? We cannot pinpoint where it exists. Since we cannot pinpoint its location, we have no idea where we should observe with the use of scientific instruments, such as infrared detectors and electron microscopes. Should we apply them to your whole body or only to your blood vessel? No answer to this question is forthcoming. Therefore, it is not a well-defined notion that actions bear moral properties.

In addition, it is not clear exactly when the action of donating blood begins to bear the property of rightness and stops bearing it. Does the action begin to assume the moral property when you visit the hospital, when you lie down on the bed, when the tip of a needle touches your skin, when the needle is pushed into your vessel, when your blood begins to flow out of your vessel, or when your blood reaches the blood bag? Does the action cease to assume the moral property when your blood stops flowing into the blood bag, when the needle is pulled out of your vessel, when the tip of the needle touches your skin, when you get up on the bed, or when you get out of the hospital? We have no idea exactly when the action of donating blood begins and ceases to bear the moral property. As a result, we have no idea exactly when we can and cannot use scientific instruments to detect it. Again, it is not a well-defined notion that actions bear moral properties.

12.3 Mathematical and Moral Fictionalism

There are many variants of mathematical and moral antirealism in the literature. In the interest of saving space, this section focuses on mathematical fictionalism (Balaguer, 2018) and moral fictionalism, which is also referred to as "error theory" (Mackie, 1977). I offer three objections that apply to both of them.

Mathematical fictionalism holds that mathematical statements purport to describe the abstract world, but that the abstract world is not real, and thus that no mathematical statement is true. The advantage of mathematical fictionalism over mathematical realism is that it does not postulate the abstract world or the faculty of reason. In consequence, it is not saddled with the burden of explicating the relationship between concreta and abstracta, does not multiply entities and faculties beyond necessity, and does not make the incredible contention that mathematical statements, although about the abstract world, can be confirmed by observations of the concrete world.

However, mathematical fictionalism is vulnerable to the following three objections.

(A) Why are some mathematical statements widely used in empirical science? On the mathematical fictionalist account, mathematics is an enterprise of describing the abstract world, which does not exist. However, empirical science is an enterprise of describing the concrete world. It is mysterious how the former is useful to the latter (Park, 2022a, p. 196). It is not explanatory for mathematical fictionalists to suggest that some mathematical statements are widely used in empirical science because they are false, for there does not seem to be any explanatory connection between the falsehood and the wide use of those mathematical statements. The suggestion only invites further questions: why are some mathematical statements widely used in empirical science despite the fact that they are false? Are false beliefs, in general, useful in empirical science?

(B) What is the difference between what laypeople take to be true and false mathematical statements, e.g., between "1 + 2 = 3" and "1 + 2 = 4"? Mathematical fictionalists cannot answer that "1 + 2 = 3" and "1 + 2 = 4" are true and false, respectively, because mathematical fictionalism asserts that no mathematical belief is true. Mathematical fictionalists end up saying that "1 + 2 = 3" and "1 + 2 = 4" are *fictionalistically correct* and *incorrect*, respectively (Balaguer, 2018). To say so means that if the abstract world existed "1 + 2 = 3" would be true while "1 + 2 = 4" would be false.

An immediate objection to this suggestion is to ask how we know whether a mathematical sentence is fictionalistically correct or incorrect. Mathematical fictionalists might answer that we know a priori that "1 + 2 = 3" and "1 + 2 = 4" are fictionalistically correct and incorrect, respectively. If they say so, however, mathematical realists would say that we know a priori that "1 + 2 = 3" and "1 + 2 = 4" correspond and do not correspond, respectively, to the abstract world. Under such circumstances, mathematical fictionalists cannot say that mathematical fictionalism is more advantageous than is mathematical realism (Park, 2022a, pp. 196–197).

(C) Finally, mathematical fictionalism violates the principle of charity, according to which we should assume that others are rational agents, which implies that we should not ceteris paribus attribute false beliefs to them (Park, 2017, p. 79). Mathematical fictionalism attributes false beliefs indiscriminately to cognitive agents holding mathematical beliefs. It does not matter whether they believe that 1 + 2 = 3 or 1 + 2 = 4. Their belief is false, according to mathematical fictionalism. There is an antirealist alternative to mathematical fictionalism that I call "mathematical inferentialism." It does not indiscriminately attribute false beliefs to cognitive agents holding mathematical beliefs, as will become clear in the next section.

Let me turn to moral fictionalism, which holds roughly that moral statements express beliefs, but that moral facts do not exist, and thus that no moral belief is true. The advantage of moral fictionalism over moral realism is that it is not committed to the existence of the abstract world or the faculty of reason. As a consequence, it is not saddled with the burden of explicating the relationship between actions and the Form of Rightness, does not multiply facts and faculties beyond necessity, and does not make the incredible contention that moral statements, although about the abstract world, can be confirmed by the observations of the concrete world.

However, moral fictionalism is vulnerable to the following three objections.

(A)' Why are some moral statements, such as "Blood donation is right" and "Stealing is wrong," widely used in our moral lives? It is not satisfactory for moral fictionalists to suggest that they are widely used because they are false, for there does not seem to be any explanatory connection between the falsehood and the wide use of those statements. The suggestion only begs further questions: why are some moral statements widely used despite the fact that they are false? Are false beliefs, in general, useful in our moral lives?

(B)' What is the difference between a moral judgment and its opposite moral judgment, e.g., between "Donating blood is right" and "Donating blood is wrong"? Moral fictionalists might answer that the former and the latter are fictionalistically correct and incorrect, respectively. To say so means that if the abstract world existed, the former would be true while the latter would be false. However, this suggestion inherits the problem with mathematical fictionalists' suggestion noted above.

(C)' Like mathematical fictionalism, moral fictionalism violates the principle of charity. It attributes false beliefs indiscriminately to cognitive agents holding moral beliefs. It does not matter whether they believe that donating blood is right or wrong. Their belief is false, according to moral fictionalism. There is an antirealist alternative to moral fictionalism that I call "moral functionalism." It does not attribute any false belief to cognitive agents holding moral beliefs, as will become clear in the next section.

12.4 Mathematical Inferentialism and Moral Functionalism

I (2017, 2022a) have developed a new form of mathematical antirealism called "mathematical inferentialism." In Chap. 11 of this book, I have developed a new form of moral antirealism called "moral functionalism." In this section, I explicate mathematical inferentialism and moral functionalism, and then argue that just as mathematical inferentialism overcomes the three objections to mathematical realism and the three objections to fictionalism, so moral functionalism overcomes the three objections to moral realism and the three objections to fictionalism unfolded in Sects. 12.2 and 12.3 above.

To put it briefly, mathematical inferentialism holds that mathematical statements expedite deductive inferences from some synthetic statements to others. Consider the following example of a deductive argument:

1 + 2 = 3
I gave you one dollar.
She gave you two dollars.
∴ You received a total of three dollars from me and her.

The first premise is a mathematical statement. The second premise, the third premise, and the conclusion are synthetic statements. The first premise enables us to see that the conclusion deductively follows from the second and third premises. As a result, the first premise facilitates the deductive inference from the second and the third premises to the conclusion. In short, a mathematical statement enables us to deductively organize synthetic statements (Park, 2022a, p. 193).

Under what conditions is a mathematical statement true? On the mathematical inferentialist account, it is true if and only if it *entails* some true synthetic statements when conjoined with other true synthetic statements. Suppose, for example, that the second premise, the third premise, and the conclusion of the argument above are true. "1 + 2 = 3" is true because it entails the true conclusion when conjoined with the two true premises (Park, 2022a, p. 194).

Under what conditions is a mathematical statement false? On the mathematical inferentialist account, it is false if and only if it entails *false* synthetic statements when it is conjoined with true synthetic statements. Consider the following example of a deductive argument:

1 + 2 = 4
I gave you one dollar.
She gave you two dollars.
∴ You received a total of four dollars from me and her.

Suppose that the second and third premises are true. "1 + 2 = 4" is false because it entails the false synthetic conclusion when conjoined with the two true synthetic premises.

How do we know whether a mathematical statement is true or false? All we need to do to know whether it is true or false is to investigate whether it entails some true or false synthetic statements when it is conjoined with other true synthetic statements, as the examples of "1 + 2 = 3" and "1 + 2 = 4" above illustrate. Thus, acquiring mathematical knowledge does not require that we grasp abstract facts; it rather requires that we should be able to make deductive inferences and observe the concrete world. Not surprisingly, mathematical inferentialism denies that the abstract world exists (Park, 2022a, p. 194). It also denies that we have the faculty of reason that purportedly grasps abstract facts.

As a corollary, none of the three objections to mathematical and moral realism in Sect. 12.2.2 above applies to mathematical inferentialism. Let me remind readers of the three objections: (i) It is not clear how a concrete object participates in (or resembles) an abstract object. Relatedly, it is odd to say that a concrete object is what it is because it participates in an abstract object. (ii) It multiplies facts and faculties beyond necessity to postulate abstract facts and the faculty of reason. (iii) It is a puzzling contention that statements about abstracta can be confirmed by observations of concreta.

In addition, none of the three objections to mathematical fictionalism in Sect. 12.3 above applies to mathematical inferentialism. The first objection to mathematical fictionalism is that it cannot explain why some mathematical statements are useful in empirical science. On the mathematical inferentialist account, they are useful in empirical science because they deductively organize synthetic statements just as a scientific theory deductively organizes observational data (Park, 2022a, p. 195).

The second objection to mathematical fictionalism is that it cannot spell out the difference between what laypeople take to be true and false mathematical statements, e.g., between "1 + 2 = 3" and "1 + 2 = 4." On the mathematical inferentialist account, the difference between them is that "1 + 2 = 3" is true while "1 + 2 = 4" is false. To say so means, as noted earlier in this section, that when conjoined with true synthetic statements, "1 + 2 = 3" entails other true synthetic statements while "1 + 2 = 4" entails false synthetic statements.

The third objection to mathematical fictionalism is that it attributes false beliefs to those who believe that 1 + 2 = 3 as well as to those who believe that 1 + 2 = 4. Unlike mathematical fictionalism, mathematical inferentialism attributes true beliefs to those who believe that 1 + 2 = 3, and false beliefs to those who believe that 1 + 2 = 4. Thus, mathematical inferentialism observes, while mathematical fictionalism violates, the principle that we should not ceteris paribus attribute false beliefs to cognitive agents (Park, 2017, p. 79).

Let me turn to moral functionalism (Sects. 11.2 and 11.5 of Chap. 11). It holds that moral utterances perform the functions of encouraging and discouraging moral actions, actions that are either right or wrong. For example, the moral judgment that donating blood is right encourages donating blood, and the moral judgment that stealing is wrong discourages stealing. Moral functionalism is noncommittal about whether moral judgments are cognitive, whether moral facts exist, and whether we know about moral facts.

Moral functionalism has the following advantages over moral realism and fictionalism. Moral functionalism does not postulate the abstract world or the faculty of reason. As a result, it is vulnerable to none of the three objections to mathematical and moral realism in Sect. 12.2 above. In addition, it is vulnerable to none of the three objections to moral fictionalism in Sect. 12.3 above. Let me remind readers of the three objections one by one and specify how moral functionalism avoids them.

The first objection to moral fictionalism is that it cannot explain why some moral statements, such as "Donating blood is right" and "Stealing is wrong," are widely used in our moral lives. On the moral functionalist account, they are widely used because they perform useful functions. "Donating blood is right" performs the useful function of encouraging potential donors to donate blood, and "Stealing is wrong" performs the useful function of discouraging potential thieves from stealing.

Let me turn to the second objection to moral fictionalism that it cannot spell out the difference between a moral statement and its opposite moral statement, e.g., between "Donating blood is right" and "Donating blood is wrong." On the moral functionalist account, the difference between them is that the former performs the useful function of encouraging potential donors to donate blood, and hence moral evaluators frequently utter it, while the latter performs the useless function of discouraging potential donors from donating blood, and hence moral evaluators seldom utter it.

The third objection to moral fictionalism is that it attributes false beliefs to those who believe that donating blood is right as well as to those who believe that donating blood is wrong. In contrast to moral fictionalism, moral functionalism is neutral regarding whether a moral judgment is a cognitive state. Accordingly, it neither attributes false beliefs to those who believe that donating blood is right, nor to those who believe that donating blood is wrong. Moral functionalism takes moral evaluators to be encouragers and discouragers, and it is noncommittal about whether they are also believers.

So far I have argued that mathematical inferentialism overcomes the three objections to mathematical realism and the three objections to mathematical fictionalism, and that moral functionalism overcomes the three objections to moral realism and the three objections to fictionalism. Mathematical and moral realists might ask the following question about mathematical inferentialism and moral functionalism. How do mathematical inferentialism and moral functionalism secure the objectivity of mathematics and morality?

Mathematical inferentialists answer that a mathematical statement is true not because we think that it is true but rather because it entails some true synthetic statements when conjoined with other true synthetic statements. The synthetic statements that the mathematical statement deductively organizes are true not because we think that they are true but rather because the world is as they say it is. In short, the truth-value of a mathematical statement is determined not by our thinking but by the logical function of entailing some true synthetic statements and by the truth-values of some synthetic statements, which are in turn determined by the concrete world (Park, 2022a, p. 195).

Moral functionalists answer that moral functionalism is neutral regarding whether morality is objective. After all, it is neutral regarding whether a moral judgment is a cognitive state, i.e., it neither claims that a moral judgment is a cognitive state, nor that it is a noncognitive state. Consequently, it neither claims that that donating blood is right independently of whether we think that donating blood is right, nor that donating blood is right simply because we feel that donating blood is right; it rather claims that moral utterances perform the functions of encouraging and discouraging moral actions. A moral utterance might be effective or ineffective in encouraging or discouraging moral actions. It is an objective matter of fact whether a moral statement is effective or ineffective in encouraging or discouraging moral actions. If a moral statement is objective at all, it is objective in that sense only.

12.5 Objection and Reply

In this section, I reply to a possible objection to mathematical inferentialism.

Mathematical inferentialism holds that a mathematical statement is true if and only if it entails some true synthetic conclusions when conjoined with true synthetic premises. Kane Baker (personal communication) provides the following counterexample to mathematical inferentialism:

$1 + 1 = 2$
You have one water droplet.
You add it to another water droplet.
∴ You have two water droplets.

The first premise is true. However, it entails the false synthetic conclusion, when conjoined with the two true synthetic premises. Mathematical inferentialists have the burden of providing an account of why it is wrong to draw the conclusion from the three premises.

This objection motivates me to perform the following thought experiment about our Paleolithic ancestors. Imagine that our Paleolithic ancestors left their two children in their cave and went out hunting and gathering. On their way back home, they saw a python slither into their cave. Before the python reached their children, they threw stones at it and drove it away from their cave. They did not think that $2 + 1 = 3$, and thus that their two children and the python would coexist peacefully in their cave. How about our ancestors' competitors? In a similar situation, they did not do anything to protect their two children from a python. Why not? They thought that $2 + 1 = 3$, and thus that their two children and the python would coexist peacefully in their cave.

This thought experiment is intended to show that the ability to use a mathematical statement correctly was an adaptive trait. Our ancestors knew, while their competitors did not, in which cases the use of a mathematical statement would lead to a true conclusion about the world, and in which other cases the use of a mathematical statement would lead to a false conclusion about the world. To state it otherwise, our ancestors knew, while their competitors did not, which synthetic statements could and could not be drawn from the conjunction of a mathematical statement and other synthetic statements. In short, our ancestors could distinguish, while their competitors could not, between correct and incorrect applications of a mathematical statement.

Objectors would ask me the following questions. What is the criterion for distinguishing between correct and incorrect applications of a mathematical statement, or between the synthetic statements that can be drawn and the synthetic statements that cannot be drawn from the conjunction of a mathematical statement and other synthetic statements? In which cases does the application of a mathematical statement lead to a true conclusion? In which other cases does the application of a mathematical statement lead to a false conclusion? Can we specify the necessary and sufficient conditions under which the use of a mathematical statement leads to true and false conclusions?

In my view, we are neither aware of the aforementioned criterion nor the aforementioned necessary and sufficient conditions. However, we can correctly use mathematical statements to draw true conclusions about the world because we use exemplars instead of a criterion or necessary and sufficient conditions. We were exposed to cases in which it was correct to use a mathematical statement to draw a true conclusion, and an exemplar case was stored in our minds. When a target case is presented to us, we compare it with the exemplar case in our minds for similarity. If the similarity between them is above a certain threshold, we use a mathematical statement to draw a conclusion; if the similarity between them is below that threshold, we do not. In short, we use an exemplar case to determine whether a mathematical statement would lead to true or false conclusions about the world. This suggestion is in line with exemplar theory (Estes, 1986), according to which concepts are stored in our minds not as necessary and sufficient conditions but as exemplars.

This exemplar account of how we recognize the cases in which a mathematical statement leads to a true conclusion can be illustrated by the following imaginary episode. Imagine that a Paleolithic woman made the following inference:

$1 + 1 = 2$
My husband has caught a fish.
My son has caught a fish.
… My family will have two fish for dinner.

This conclusion turned out to be true. Such an inference was stored in her mind. It served as an exemplar inference with which she compared a target inference for similarity. She was asked whether she would make the following inference:

$1 + 1 = 2$
You have one water droplet.
You add it to another water droplet.
… You have two water droplets.

She refused to make this inference on the grounds that it was not sufficiently similar to the exemplar inference in her mind. She immediately knew that "1 + 1 = 2" cannot be used to draw the conclusion that you would have two water droplets.

What philosophical lesson can we draw from the foregoing imaginary episode? Our ability to use a mathematical statement correctly can be explained by our ability to make similarity judgments between target and exemplar inferences. We do not need to invoke a criterion for distinguishing between correct and incorrect applications of a mathematical statement. Nor do we need to invoke the necessary and sufficient conditions under which a mathematical statement leads to true and false conclusions.

12.6 Conclusion

Just as mathematical inferentialism circumvents the three objections to mathematical realism and the three objections to mathematical fictionalism, so moral functionalism circumvents the three objections to moral realism and the three objections to moral fictionalism. Mathematical realism is so problematic that moral realists have nothing to gain from comparing moral statements to mathematical statements. To compare moral statements to mathematical statements is, if anything, to make moral statements unbelievable. This chapter has the following simple message to moral realists who appeal to mathematical realism: "Mathematical realism doesn't help moral realism."

References

Baker, A. (2005). Are there genuine mathematical explanations of physical phenomena? *Mind, 114*(454), 223–237.

Baker, A. (2009). Mathematical explanation in science. *The British Journal for the Philosophy of Science, 60*(3), 611–633.

Balaguer, M. (2018). Fictionalism in the philosophy of mathematics. In E. N. Zalta (Ed.), *The Stanford encyclopedia of philosophy*. https://plato.stanford.edu/archives/fall2018/entries/fictionalism-mathematics/

Bourget, D., & Chalmers, D. (2014). What do philosophers believe? *Philosophical Studies, 170*(3), 465–500.

DeLapp, K. (2023). Metaethics. *The Internet encyclopedia of philosophy*. https://iep.utm.edu/metaethi/

Estes, W. (1986). Array models for category learning. *Cognitive Psychology, 18*(4), 500–549.

Huemer, M. (2016). A liberal realist answer to debunking skeptics: The empirical case for realism. *Philosophical Studies, 173*(7), 1983–2010.

Hume, D. (1978). In L. A. Selby-Bigge & P. H. Nidditch (Eds.), *A treatise of human nature*. Oxford University Press.

Jones, T. (1970). *The classical mind: A history of Western philosophy*. Harcourt Brace Jovanovich College Publishers.

Kim, S. (2023). Moral realism. *The internet encyclopedia of philosophy*. https://iep.utm.edu/moralrea/

Mackie, J. (1977). *Ethics: Inventing right and wrong*. Penguin Books.

Park, S. (2003). Ontological order in scientific explanation. *Philosophical Papers, 32*(2), 157–170.

Park, S. (2017). In defense of mathematical inferentialism. *Analysis and Metaphysics, 16*, 70–83.

Park, S. (2022a). *Embracing scientific realism*. Springer.

Park, S. (2022b). Moral subjectivism vs. moral objectivism. *Filosofija. Sociologija, 33*, 191–198.

Putnam, H. (1971). *Philosophy of logic*. Harper and Row.

Quine, W. (1948). On what there is. *Review of Metaphysics, 2*(5), 21–38.

Chapter 13
Metaethical and Non-metaethical Explanations of Moral Progress

Abstract Some moral realists suggest that morality has progressed because some individuals recognized moral facts which are causally inert. I object causally inert facts, whether moral or mathematical, cannot serve as explanantia for anything, and then provide non-metaethical and moral antirealist explanations of why morality has progressed. I argue that moral antirealists, who aim to undermine the moral realist explanation, should provide a moral antirealist explanation rather than a non-metaethical explanation of why morality has progressed. I also point out that metaethical and non-metaethical explanations have respective limitations.

Keywords Huemer · Metaethics · Moral antirealism · Moral facts · Moral judgments · Moral progress · Moral realism

13.1 Introduction

Morality has progressed steadily for millennia and significantly over the past 200 years (Huemer, 2016, pp. 1988–1993). During this period, for example, the murder rate has dropped significantly. Conquerors came to be no longer admired. Slavery and gruesome punishments were abolished. Women were no longer executed for committing adultery. Dictatorial governments were replaced with democratic ones. Colonies achieved independence from imperial countries. In short, our modern ancestors became morally better than our Paleolithic ancestors.

Why did our modern ancestors become morally better? There can be diverse explanations of why morality has progressed, and the diverse explanations can be divided into metaethical and non-metaethical ones. Metaethical explanations invoke moral explanantia, such as moral facts and judgments, while non-metaethical explanations invoke non-moral explanantia, such as the influence of science and

S. Park, *Induction, Science, and Morality*, Synthese Library 520,
https://doi.org/10.1007/978-3-032-00032-3_13

technology and the instinctive aversion to incest, as will become clear below. Metaethical explanations can, in turn, be divided into moral realist and antirealist ones. The moral realist explanation invokes both moral facts and judgments, while moral antirealist explanations invoke moral judgments, but not moral facts.

This chapter has several aims: to criticize the moral realist explanation, to criticize moral antirealists' critical response to the moral realist explanation, to shed light on the relationship between metaethical and non-metaethical explanations, to develop moral antirealist explanations which I take to be genuine alternatives to the moral realist explanation, and to point out the limitations of both metaethical and non-metaethical explanations.

The outline of the present chapter is as follows. In Sect. 13.2, I critically assess Michael Huemer's (2016) moral realist explanation, according to which our modern ancestors became morally better than our Paleolithic ancestors because some of them recognized moral facts. In Sect. 13.3, I develop a non-metaethical explanation that I call "the sci-tech explanation," according to which our modern ancestors became morally better than our Paleolithic ancestors because the influence of science and technology became strong on their moral lives. In Sect. 13.4, I argue that evolutionary theory in conjunction with other hypotheses can explain why our modern ancestors became morally better than our Paleolithic ancestors. In Sect. 13.5, I argue that moral antirealists cannot rest content with the sci-tech explanation, and that they should develop moral antirealist explanations, the metaethical explanations that invoke moral judgments, but not moral facts. In Sect. 13.6, I unfold moral antirealist explanations, the genuine rivals to Huemer's moral realist explanation. In Sect. 13.7, I shed light on the limitations of metaethical and non-metaethical explanations from the perspective of Bas van Fraassen's (1980) contextual theory of explanation. In Sect. 13.8, I present a new research topic for future metaethicists.

This chapter is intended to be useful to those who wonder why our modern ancestors became morally better, how metaethical explanations relate to non-metaethical explanations of moral progress, and in what contexts metaethical and non-metaethical explanations are appropriate.

13.2 The Moral Realist Explanation and Its Problems

In this section, I explicate Huemer's moral realist explanation of why morality has progressed and then critically assess it.

Huemer states that "certain moral facts explain the phenomenon of liberalization over the course of human history" (Huemer, 2016, p. 2006). For example, the moral fact that slavery was wrong explains "why slavery was abolished" (Huemer, 2016, p. 2006). Some individuals whose faculty of reason was more powerful than those of other individuals recognized this moral fact, and their moral belief that slavery was wrong spread through their culture. As a result, slavery was abolished. Huemer writes, "What is necessary is only that, at a given point in time, there be some individuals who are capable of seeing certain moral issues a bit more accurately than most of their contemporaries" (Huemer, 2016, p. 2005).

Huemer opines that moral "facts are causally inert," but that we have the "capacity to apprehend moral truths" (Huemer, 2016, p. 2006). In other words, we have the faculty of reason that enables us to recognize moral fact, which are causally inert. His view that we have the faculty of reason "seemed plausible to many realists, from Plato to W. D. Ross" (Huemer, 2016, pp. 2003–2004). On Plato's account, we can have mathematical knowledge, although "abstract, mathematical objects do not cause anything to happen" (Huemer, 2016, p. 2006). Huemer also says that "our moral knowledge is of a kind with our other a priori knowledge, such as our knowledge of mathematics and of necessary truths of metaphysics" (Huemer, 2016, p. 1986). For example, "the knowledge that pain is bad is a priori" (Huemer, 2016, p. 1986, footnote).

How does Huemer's account of moral knowledge relate to Alan Baker's (2005, 2009) account of mathematical knowledge? According to the latter, some mathematical facts explain concrete phenomena, and thus we have knowledge about mathematical facts. If inference to the best explanation (IBE) establishes concrete hypotheses, it can also establish mathematical hypotheses, which implies that if experience justifies concrete hypotheses, it can also justify mathematical hypotheses. Huemer can similarly suggest that moral facts explain moral progress, and that if IBE establishes concrete hypotheses, it can also establish moral hypotheses.

However, Huemer does not suggest so; he rather suggests that some moral knowledge is a priori. For example, "the knowledge that pain is bad is a priori" (Huemer, 2016, p. 1986, footnote). In my view, this example is disputable. If pain were a priori bad, it would be a priori bad to punish criminals. However, it is a posteriori good to punish criminals. Thus, some pain is a posteriori good. In addition, the capacity to feel pain is an adaptive trait, given that pain brings the mind's attention to bodily damage, thereby prodding the mind to take action regarding the bodily damage.

In response, Huemer might provide different examples: "Unnecessary suffering is wrong" and "Equals should be treated equally" (Schick & Vaughn, 2010, p. 365). These moral statements do not admit of counterexamples, and they are, Huemer might continue, a priori. What are we to make of those moral statements? My answer is that they are analytic (Park, 2014, p. 50). They are true in virtue of what it means for suffering to be unnecessary and what it means for individuals to be equal. They contain no information about human actions, and hence they do not count as moral statements in the first place. I challenge moral realists, those who believe that some moral knowledge is a priori, to provide an example of such knowledge.

There is an internal conflict in Huemer's position. On the one hand, he claims that moral facts are causally inert. On the other hand, he claims that slavery was abolished because some individuals recognized the moral fact that slavery was wrong and formed the moral belief that slavery was wrong. It is inscrutable to me how some individuals could identify that moral fact when it was causally inert. To say that some individuals recognized the moral fact and formed the moral belief means that the moral fact had an impact on their minds, i.e., that the moral fact causally affected their minds. It is to have it both ways to say that moral facts are causally inert, but that some individuals identified them.

This criticism of Huemer's account of moral knowledge applies no less to mathematical realists' account of mathematical knowledge, according to which we have mathematical knowledge about mathematical facts although they are causally inert. In consequence, it is not a convincing response to say that just as we have mathematical knowledge about mathematical facts which are causally inert, so we can have knowledge about moral facts. Let me remind readers that I argued in Chap. 12 that mathematical realism does not help moral realism at all.

As noted earlier in this section, Huemer states that moral knowledge is similar to mathematical knowledge (Huemer, 2016, p. 1986). However, he does not directly state whether moral facts are abstract or concrete. As such, I construct the following dilemma against him. On the one hand, if he claims that moral facts are abstract, his position is liable to all the criticisms that I leveled against the abstract version of moral realism in Chap. 12, which I concluded with the slogan: "Mathematical realism doesn't help moral realism." To mention just one of those criticisms, it is to put the cart before the horse to say that a concrete object is what it is because it resembles (or participates in) an abstract object. On the other hand, if Huemer claims that moral facts are concrete, his position is liable to all the criticisms that I leveled against the concrete version of moral realism in Sects. 10.2, 10.4, 10.5, and 10.6 of Chap. 10, in Sect. 11.4 of Chap. 11, and in Sect. 12.2.3 of Chap. 12. For example, it is not a well-defined notion that actions bear moral properties.

Let me add here that if Huemer claims that moral facts are concrete, his position is subject to the following additional objection. It is to have it both ways to say that moral facts are concrete and causally inert. Consider an orange on the food table. The orange can exist because it interacts with other objects, including the food table. It cannot exist without interacting with other objects in the concrete world. I challenge moral realists to provide an example of a fact other than a moral fact that can exist without interacting with other facts in the concrete world. As it stands, it is unintelligible to say that moral facts, although causally inert, can exist in the concrete world.

Why do moral realists strive to explain moral progress? What do they aim to achieve with the moral realist explanation of moral progress? They would answer that they aim to secure the objectivity of moral judgments. The idea is that if moral facts existed, the truth-values of moral judgments would be objective, i.e., moral judgments would be true or false not because we think so but rather because moral facts make or do not make them so. In short, if moral facts explain moral progress, we can believe that they exist, and hence that truth-values of moral judgments are objective.

In my view, however, this line of reasoning requires an additional premise that moral facts are causally efficacious, i.e., without the additional premise, the line of reasoning is disputable. To state it otherwise, even if moral facts best explain moral progress, it is questionable whether they exist, insofar as they are alleged to be causally inert. The alleged causal inertness of moral facts puts a damper on moral realists' aspiration to secure the objectivity of truth-values of moral judgments.

Moral realists might retort that if IBE can establish scientific hypotheses, it can also establish moral hypotheses. In my view, however, this retort can be countered by the consideration of a relevant difference between scientific and moral explanations. Namely, a scientific explanation captures, while a moral explanation does not, a causal relationship. Consider, for example, the scientific hypothesis that marsupials lived in Antarctica millions of years ago. It explains why marsupial fossils exist in Antarctica. The explanation captures the causal relationship between the explanans and the explanandum. Accordingly, the existence of the explanandum establishes the existence of the explanans. By contrast, consider the moral hypothesis that moral facts existed. It allegedly explains why our modern ancestors became morally better. However, there is no causal relationship between the explanans and the explanandum. Consequently, the existence of the explanandum does not establish the existence of the explanans.

To put it in general terms, in cases where there is a causal relationship between an explanans and an explanandum, IBE has the epistemic force to establish the existence of the explanans. By contrast, in cases where there is no causal relationship between them, IBE has no epistemic force to establish the existence of the explanans. In my view, to say that moral and mathematical facts are causally inert is to say that their existence cannot be established no matter how well they may explain phenomena.

To go further, it is even questionable whether a causally inert thing can explain anything. Imagine that someone says to you that she is happy because she has a cat, i.e., that the existence of the cat explains why she is happy. To your surprise, however, she adds that the cat is causally inert! You would wonder what the word "because" means in her sentence "I am happy because I have the cat," what the word "explains" means in her sentence "The existence of the cat explains why I am happy," and what justifies the attachment of the label "explain" to the relationship between the cat and her happiness.

By parity of reasoning, you should wonder what the word "explain" means in Huemer's sentence that "certain moral facts explain the phenomenon of liberalization over the course of human history" (Huemer, 2016, p. 2006), and what justifies the attachment of the label "explain" to the relationship between moral facts and moral progress. It appears to me that just as it is unjustifiable to say that the existence of the cat, which is causally inert, explains why someone is happy, so it is unjustifiable to say that a moral fact, which is causally inert, explains why slavery was abolished.

My objection to Huemer's position can be put differently. Consider his contention that moral facts explain moral progress, i.e., that our ancestors became morally better because moral facts existed. This explanation is built upon the counterfactual conditional that if moral facts had not existed, our ancestors would not have become morally better. This counterfactual conditional implies that moral facts were a cause and moral progress was an effect, which conflicts with Huemer's another contention that moral "facts are causally inert" (Huemer, 2016, p. 2006). In consequence, it is to have it both ways for Huemer to say that moral facts are causally inert but explain moral progress.

The same objection can be raised mutatis mutandis to mathematical realists. It is to have it both ways for mathematical realists to say that mathematical facts are causally inert but explain concrete phenomena. Therefore, it is not convincing for moral realists to say that just as mathematical facts, although causally inert, explain concrete phenomena, so moral facts, although causally inert, explain moral progress.

Huemer might object that my complaint relies on the causal theory of explanation (Salmon, 1978), but that there are other theories of explanation, such as the deductive-nomological model (Hempel, 1966), the unification theory (Friedman, 1974; Kitcher, 1981), and the contextual theory (van Fraassen, 1980). There might be a lawlike relationship between moral facts and moral progress, in which case his explanation of moral progress fits the deductive-nomological model. Moral facts might unify diverse moral phenomena including moral progress, in which case his explanation of moral progress fits the unification theory. These possible suggestions are interesting to me. Unfortunately, it requires digression to develop and refute them. Suffice it to say that I am not sanguine about their prospects, and that I argue at length in Sect. 13.7 below that Huemer's explanation of moral progress does not fit the contextual theory.

Let me raise another objection to Huemer's moral realist explanation of moral progress. Why was slavery abolished about 200 years ago? Why not about 2000 years ago? How could some individuals grasp the moral fact that slavery was wrong about 200 years ago, but not about 2000 years ago? It is not plausible for Huemer to answer that some individuals acquired the more powerful faculty of reason about 200 years ago, and that as a result they could apprehend the moral fact that slavery was wrong.

This criticism of Huemer's moral realist explanation parallels his criticism of the evolutionary explanation for why slavery was abolished about 200 years ago. The evolutionary explanation holds, according to Huemer, that slavery was abolished because our modern ancestors became morally better as a result of a genetic change that occurred in them about 200 years ago. Huemer objects that this evolutionary explanation is "extremely implausible empirically," and hence that evolutionary theory "does not account for the *recency* of liberal values" (Huemer, 2016, p. 1995). In my view, his criticism against the evolutionary explanation can be turned against his moral realist explanation.

13.3 The Sci-Tech Explanation

In this section, I develop what I call "the sci-tech explanation," according to which morality has progressed thanks to the advancement of science and technology.

Modern science has improved our worldviews and our moral lives. For example, before the advent of modern science, some religious people regarded the sun as a supernatural entity, and they sacrificed humans to worship it. Some people thought that witches were real, and they burned many innocent women to death (Huemer, 2016, p. 1990). Modern science made such religious beliefs less influential on our forebears' moral lives. In general, what was once considered (morally) right can be considered as (morally) wrong with the advancement of science.

Modern technology has also improved our forebears' moral lives. For example, thousands of years ago, many women were executed for committing adultery (Huemer, 2016, p. 1998). The Bible, which was written about 2000 years ago, says that if a bride is not a virgin on the wedding night, "the men of her town shall stone her to death" (Deuteronomy 22:21). Such practices are no longer acceptable today. In my view, this moral change was brought about in part thanks to the invention of contraception. Premarital sex and adultery could have resulted in unintended pregnancy in ancient times. With the invention of contraception, however, unintended pregnancy could be effectively prevented. As a result, casual sex became permissible. In general, what was once considered wrong can be considered permissible with the advancement of technology.

Let me move on to democratization. Dictatorial governments were replaced with democratic ones in the modern era (Huemer, 2016, pp. 1992–1993). In ancient times, for example, kings enjoyed legislative, executive, and judiciary powers. Moreover, they were in power for life and transferred their powers to their sons. Such practices have been put to an end. An interesting question arises. What contribution did science and technology make to the replacements of dictatorial governments with democratic ones?

Our modern ancestors could form unions against their dictatorial rulers with the advancement of science and technology. The modern means of communication, such as postal mail, printing techniques, the telegraph, and the phones, enabled them to share their antipathy toward their dictatorial rulers. The modern means of transportation, such as trains, cars, and airplanes, enabled them to gather together and to raise their voice together against their dictatorial rulers. As a result, dictatorial political systems were replaced with democratic ones. In short, science and technology contributed to democratization.

We cannot even imagine that a ruler of a country today declares that she will stay in power for life, that rebels should be dismembered, that men can have concubines, that slaves should obey masters, that torture is a legitimate interrogation method, that women should be disallowed to vote, that deformed babies should be killed, and that servants should be buried alive with dead rulers. If a ruler seriously declared so, she would only be expelled from power by her people and by the people of other countries who cherish human rights. Such an expulsion would be impossible without modern conveniences, such as television, the Internet, the telephone, vehicles, aircraft, and weapons.

13.4 The Conjunction of Evolutionary Theory and Other Hypotheses

Huemer (2016, pp. 1994–1996) contends, as noted earlier, that evolutionary theory cannot explain why slavery was abolished about 200 years ago because it is extremely implausible to say that our modern ancestors became morally better as a result of a genetic change that occurred in them about 200 years ago. In this section,

I argue that evolutionary theory in conjunction with other hypotheses can explain moral progress.

Our Paleolithic ancestors traveled around in small groups in search of animals and fruits. The members of a small group that traveled together might be called "in-group members," and the members of another group that traveled separately might be called "out-group members," to use the terms in evolutionary biology. In-group members were family members, relatives, and acquaintances to one another. They interacted with one another on daily basis, but not with out-group members.

Imagine that there were four kinds of humans in the Paleolithic Age. The first kind of humans were friendly to in-group members but unfriendly to out-group members. The second kind of humans were equally friendly to in-group and out-group members. The third kind of humans were equally unfriendly to in-group and out-group members. The fourth kind of humans were unfriendly to in-group members but friendly to out-group members. These four kinds of humans competed with one another for survival and reproduction in the Paleolithic Age. Who would have persisted? Who would have perished?

Answers to these questions can be found in the following thought experiment. In the Paleolithic Age, traveling groups sometimes encountered one another. The four kinds of humans characterized above reacted differently to out-group members. The first and third kinds of humans deprived out-group members of their means of survival and reproduction, and hence they were rewarded by in-group members. The second and fourth kinds of humans did not do any harm to out-group members. When a traveling group did not encounter another traveling group, the first and second kinds of humans were friendly to in-group members, and hence they were rewarded by in-group members. By contrast, the third and fourth kinds of humans were unfriendly to in-group members, and hence they were punished by in-group members. Obviously, the first kind of humans were more likely to survive and reproduce than any other kinds of humans. The first kind of humans had children, and their children inherited the inclination to be friendly to in-group members and unfriendly to out-group members. After many generations, the first kind of humans persisted while the other three kinds of humans perished.

This speculative story is intended to show that the inclination to be friendly to in-group members and unfriendly to out-group members was an adaptive trait in the Paleolithic Age (Rusch, 2014, pp. 1–2), that we inherited the inclination from our Paleolithic ancestors, and that we are genetically hard-wired to be friendly to in-group members and unfriendly to out-group members. This hypothesis ties in with an anthropological study of a tribe in South America (Chagnon, 1988, p. 989), according to which warriors have 1.62 wives and 4.91 children on average while nonwarriors have 0.63 wives and 1.59 children on average. Note that those who are unfriendly to out-group members are reproductively more successful than those who are not unfriendly to out-group members.

There is a crucial difference between the Paleolithic Age and the modern era. In the Paleolithic Age, there was a clear line between in-group and out-group members, given that in-group members were those who traveled around together. In the modern era, however, there was no such boundary. Our modern ancestors lived in permanent places where they grew crops and raised livestock. Those who lived nearby were not out-group members but neighbors. Neighbors had two faces: those of in-group and out-group members, i.e., those of friends and enemies. In short, our modern ancestors were more friendly to out-group members than were our Paleolithic ancestors. Let me call "the settlement hypothesis" the idea that the lifestyle of living in permanent places contributed to moral progress.

There is another crucial difference between the Paleolithic Age and the modern era. In the modern era, our ancestors had modern science and technology. Modern science and technology enabled them to interact with their neighbors effectively and frequently. Furthermore, modern science and technology pushed the boundaries of neighbors to include more people as neighbors. Many people, although physically far from one another, became neighbors thanks to the efficient means of communication and transportation, such as postal mail, printing techniques, the telegraph, the phones, television, the Internet, trains, cars, and airplanes.

Why were our modern ancestors morally better than our Paleolithic ancestors? An answer to this question can be inferred from the conjunction of evolutionary theory, the settlement hypothesis, and the sci-tech hypothesis. To put it briefly, our modern ancestors inherited the propensity to be friendly to in-group members from our Paleolithic ancestors. Unlike our Paleolithic ancestors, our modern ancestors lived in permanent places, and they treated those who lived nearby as neighbors rather than as out-group members. In addition, they were equipped with modern science and technology. As a result, they became morally better than our Paleolithic ancestors. This explanation of moral progress does not require that a genetic mutation occurred in our modern ancestors; it rather requires that they were genetically identical with our Paleolithic ancestors.

13.5 Metaethical and Non-metaethical Explanations

In this section, I point out that Huemer's moral realist explanation is a metaethical one while the sci-tech explanation is a non-metaethical one, and then argue that moral antirealists cannot rest content with the sci-tech explanation.

Let me first summarize Huemer's moral realist explanation and the sci-tech explanation. The moral realist explanation holds that our recent ancestors became morally better because some individuals recognized moral facts. By contrast, the sci-tech explanation holds that our recent ancestors became morally better because the influence of science and technology on their moral lives became strong.

Both the moral realist and sci-tech explanations can predict that our descendants will become morally better than are we. However, they require different initial conditions to make the same prediction. The moral realist explanation requires the initial condition that some individuals will be born in the future with the more powerful faculty of reason that will be more sensitive to moral facts. In contrast, the sci-tech explanation requires the initial condition that our descendants will have more advanced science and technology. With more advanced science and technology, they will interact with one another more effectively and frequently. As a result, they will be more friendly to each other.

How would moral antirealists respond to the sci-tech explanation? They would help themselves to it, given that it entitles them to say that we do not need to invoke moral facts to explain why our recent forebears became morally better. They might add that the sci-tech explanation has the following two advantages over the moral realist explanation. First, the moral realist explanation multiplies facts and cognitive faculties beyond necessity while the sci-tech explanation does not. After all, the moral realist explanation postulates, while the sci-tech explanation does not, the faculty of reason that purportedly apprehends moral facts. Second, the sci-tech explanation has an adequate answer, while the moral realist explanation does not, to the following question. Why did moral progress accelerate about 200 years ago as opposed to about 2000 years ago? It is plausible to say that moral progress accelerated about 200 years ago because the influence of modern science and technology on our ancestors' moral lives accelerated about 200 years ago. By contrast, it is implausible to say that moral progress accelerated about 200 years ago because our recent forebears acquired the more powerful faculty of reason as a result of genetic mutation.

How would moral realists respond to the sci-tech explanation? Like moral antirealists, they would help themselves to it, given that it can enrich their explanation of moral progress. On their explanation, our recent ancestors became morally better because some individuals recognized the moral fact that slavery was wrong and their moral belief that slavery was wrong spread to their culture. How could their moral belief that slavery was wrong spread to their culture? According to the sci-tech hypothesis, their moral belief spread to their culture expeditiously thanks to the efficient means of communication, such as printing techniques and the telegraph.

As just noted, both moral realists and antirealists can help themselves to the sci-tech explanation, which is not surprising, given that the sci-tech explanation neither says that moral facts exist nor that they do not. The neutrality of the sci-tech explanation between moral realism and antirealism brings our attention to the fundamental difference between the moral realist and sci-tech explanations. The moral realist explanation is a metaethical one in that it invokes moral explanantia—moral facts and judgments—to explain moral progress. By contrast, the sci-tech explanation is a non-metaethical one in that it does not invoke moral explanantia.

The sci-tech explanation is not the only non-metaethical explanation of moral progress, and there are other non-metaethical explanations of moral progress. Let me sketch just one of them that I call "the reproductive explanation." Imagine that there were two kinds of men in the Paleolithic Age. The first group of men had an instinctive aversion to incest, while the second group of men did not. The first kind of men attacked other groups and killed all the out-group members except virgins, with whom they had children together.[1] By contrast, the second kind of men were

[1] This assumption receives support from the Bible. In the Bible, Moses orders his military officers to massacre all Midianites except virgins, saying that they should "kill all the boys, as well as every

not interested in out-group virgins but in-group virgins. The first kind of men were more likely to have healthy babies than the second kind of men. The capacity to feel an aversion to incest was passed on to future generations. After many generations, the first kind of men persisted while the second kind of men perished.

With the advent of the New Stone Age, our ancestors began agriculture and lived in permanent places. Neighboring groups of people regarded one another as neighbors with whom they could mate. Children were born from men and women of different groups, and they were friendly both to their fathers' groups and mothers' groups. As a result, the members of the different groups became more friendly to one another. In short, our instinctive aversion to incest was one of the factors that contributed to moral progress. This explanation of moral progress might be called "the reproductive explanation."

The sci-tech and reproductive explanations entitle moral antirealists to say that we do not need to invoke moral facts to explain moral progress. In my view, however, moral antirealists cannot rest content with them, given that moral antirealists are neither historians of science and technology nor historians of reproductive patterns but rather metaethicists. Metaethicists are those who are interested in what moral judgments are, in what role moral judgments play in our moral lives, and in what contributions moral judgments make to social changes. In consequence, they should explain moral progress in terms of moral judgments. In the next section, I explore how moral antirealists can meet this burden of explanation.

13.6 Moral Antirealist Explanations

There are a variety of versions of moral antirealism. In this section, I unpack their diverse explanations of moral progress.

Emotivism (Stevenson, 1944) holds that a moral judgment is the expression of an emotion. For example, moral evaluators'[2] judgment that stealing is wrong is their negative emotion toward stealing. On the emotivist account, slavery was abolished because our modern ancestors expressed a negative emotion toward slavery. Why did they express a negative emotion toward slavery? They inherited the propensity to be friendly to in-group members from our Paleolithic ancestors. Unlike our Paleolithic ancestors, they lived in permanent places and were armed with modern science and technology. As a result, the distinction between in-group and out-group members became blurred, and they treated those who lived nearby like neighbors rather than like out-group members. In consequence, they treated slaves as their neighbors, and expressed a negative emotion toward slavery. Note that emotivists explain the abolition of slavery in terms of the expression of the negative emotion,

woman who has had relations with a man, but spare for yourselves every girl who has never had relations with a man" (Numbers 31:17–18).

[2] Moral evaluators are those who make moral judgments.

and then explain the expression of the negative emotion in terms of the conjunction of evolutionary theory, the settlement hypothesis, and the sci-tech hypothesis.

Prescriptivism (Hare, 1952) holds that a moral judgment is a command. For example, moral evaluators' moral judgment that stealing is wrong is a command "We ought not to steal." On the prescriptivist account, slavery was abolished because our modern ancestors expressed the command "Slavery ought to be abolished." Why did they express the command? Like emotivists, prescriptivists can appeal to the conjunction of evolutionary theory, the settlement hypothesis, and the sci-tech hypothesis to answer this question.

Error theory (Mackie, 1977) holds that moral judgments are truth-apt, that moral properties, being queer, do not exist, and that no moral judgment is true. Moral evaluators' moral judgment that stealing is wrong is a moral belief that stealing is wrong, but the belief is false because there is no moral fact that would make it true. On the error theorist account, slavery was abolished because our modern ancestors expressed the false belief that slavery was wrong. Unfortunately, this explanation does not sound convincing. It only begs the following questions. Why was slavery abolished despite the fact that our modern ancestors' belief was false? Do false beliefs, in general, contribute to amazing events, such as the abolition of slavery?

Why did our modern ancestors express the false belief that slavery was wrong? Error theorists would provide the following answer. Our modern ancestors inherited the propensity to be friendly to in-group members from our Paleolithic ancestors. Unlike our Paleolithic ancestors, they lived in permanent places and were equipped with modern science and technology. As a result, they expressed the false belief.

Unfortunately, this explanation does not sound convincing, given that there are no explanatory connections between the propensity to be friendly to in-group members and the propensity to express false beliefs, between the lifestyle of living in permanent places and the propensity to express false beliefs, and between the advancement of science and technology and the propensity to express false beliefs. It does not sound convincing to say, for example, that we have the propensity to express false beliefs when we live in permanent places.

Quasi-realism (Blackburn, 1993) holds that a moral judgment is not cognitive, but that we can say that it is true because saying that it is true that p is equivalent to saying that p. Moral evaluators' moral judgment that stealing is wrong is a negative emotion toward stealing, but they can say that it is true that stealing is wrong. On the quasi-realist account, slavery was abolished because moral evaluators expressed a negative emotion toward slavery. Why did they express a negative emotion toward slavery? Like emotivists, quasi-realists would appeal to the conjunction of evolutionary theory, the settlement hypothesis, and the sci-tech hypothesis to answer this question.

Moral skepticism (Street, 2006) holds that no moral judgment is justified. On the moral skeptical account, slavery was abolished because our modern ancestors expressed the unjustified belief that slavery was wrong. Unfortunately, this explanation does not sound convincing. There is no explanatory connection between the expression of the unjustified belief and the abolition of slavery. The moral skeptical explanation invites the following questions. Why was slavery abolished despite the fact that our modern ancestors' belief about slavery was not justified? Do unjustified beliefs, in general, contribute to amazing events, such as the abolition of slavery?

Why did our modern ancestors express the unjustified belief? Moral skeptics would provide the following answer. Our modern ancestors inherited the propensity to be friendly to in-group members from our Paleolithic ancestors. Unlike our Paleolithic ancestors, they lived in permanent places and were equipped with modern science and technology. As a result, they expressed the unjustified belief.

Unfortunately, this explanation does not sound convincing, given that there are no explanatory connections between the propensity to be friendly to in-group members and the propensity to express unjustified beliefs, between the lifestyle of living in permanent places and the propensity to express unjustified beliefs, and between the advancement of science and technology and the propensity to express unjustified beliefs. It does not sound convincing to say, for example, that we have the propensity to express unjustified beliefs when we have the advancement of science and technology.

Cultural emotivism (Sect. 10.3 of Chap. 10) asserts that a moral statement is a description of cultural emotion. Moral evaluators' moral judgment that stealing is wrong is the belief that their culture has a negative cultural emotion toward stealing. On the cultural emotivist account, slavery was abolished because moral evaluators expressed the belief that their culture had a negative emotion toward slavery. Why did they express the belief? Like emotivists, cultural emotivists can appeal to the conjunction of evolutionary theory, the settlement hypothesis, and the sci-tech hypothesis to answer this question.

Moral functionalism (Sects. 11.2 and 11.5 of Chap. 11) asserts that moral utterances perform the functions of encouraging and discouraging moral actions, actions that are either right or wrong. It is neutral regarding whether a moral judgment is a cognitive state, and whether moral properties are real. For example, the moral judgment that charitable giving is right encourages charitable giving, and the moral judgment that stealing is wrong discourages stealing. On the moral functionalist account, slavery was abolished because our modern ancestors' utterance "Slavery is wrong" performed the function of discouraging slavery. Why did they express the moral judgment that performed the function of discouraging slavery? Like emotivists, moral functionalists can appeal to the conjunction of evolutionary theory, the settlement hypothesis, and the sci-tech hypothesis to answer this question.

All these explanations for the abolition of slavery invoke moral judgments. Hence, they are all metaethical explanations, and hence they are genuine competitors to the moral realist explanation. As a consequence, moral antirealists are required to show that their metaethical explanations are better than the moral realist explanation, but not that non-metaethical explanations, such as the sci-tech and reproductive explanations, are better than the moral realist explanation. To compare a non-metaethical explanation with the moral realist explanation is to compare an apple with an orange.

The moral antirealist explanations sketched above have the following two advantages over the moral realist explanation. First, none of the moral antirealist explanations claims that there are moral facts, and that we have the faculty of reason that can discover them. Thus, in contrast to the moral realist explanation, they do not multiply facts and faculties beyond necessity. Second, the moral antirealist explanations assert that there is a causal relationship between their explanantia and the explanandum. For example, the emotivist explanation asserts that our modern ancestors' expression of the negative emotion toward slavery contributed to the abolition of slavery. As a consequence, moral antirealist explanations are not vulnerable to my criticisms to the moral realist explanation in Sect. 13.2 above, e.g., IBE cannot establish the existence of a causally inert explanans.

13.7 Relevant Factors

In the previous sections, I introduced metaethical and non-metaethical explanations of why morality has progressed. In this section, I cast light on how they relate to each other from the perspective of van Fraassen's (1980) contextual theory of explanation.

The contextual theory holds that "An explanation is an answer to a why-question" (van Fraassen, 1980, p. 134), and "among the scientifically relevant factors, context determines explanatorily relevant ones" (van Fraassen, 1980, p. 126). Suppose that many *factors*—F_1, F_2, and so on—jointly produced an event. The questioner asks the explainer to explain why the event occurred. On the contextual theory, the explainer should choose a relevant factor from the set of factors that jointly produced the event, and context determines which factor is a relevant one. F_1 is a relevant factor in the context where the questioner is interested in whether F_1 was present. F_2 is a relevant factor in the context where the questioner is interested in whether F_2 was present, and so forth.

We can apply the contextual theory to the explanation of why our modern ancestors became morally better. Many factors jointly produced moral progress, including the inclination to be friendly to in-group members, the lifestyle of living in permanent places, the influence of science and technology, the influence of the instinctive aversion to incest, and the influence of moral judgments. The influence of science and technology is a relevant factor in the context where the questioner is interested in whether science and technology contributed to moral progress, but not in other contexts. The influence of moral judgments is a relevant factor in the context where the questioner is interested in whether moral judgments contributed to moral progress, but not in other contexts.

So what? Metaethical and non-metaethical explanations have their respective limitations. Metaethical explanations are appropriate to metaethicists who are interested in whether moral factors, such as moral facts and judgments, contributed to moral progress, but not to non-metaethicists who are interested in whether non-moral factors, such as science and technology, contributed to moral progress. The same is true of non-metaethical explanations. They are appropriate to non-metaethicists who are interested in whether non-moral factors contributed to moral progress, but not to metaethicists who are interested in whether moral factors contributed to moral progress.

The moral realist, error theorist, and moral skeptical explanations do not fit the contextual theory. Moral facts, the ultimate explanantia that moral realists invoke, are causally inert. Causally inert facts cannot be *factors* for anything. As a consequence, it is not clear in what context the moral realist explanation would be an appropriate answer. How about error theory and moral skepticism? As noted in Sect. 13.6 above, it does not sound convincing to say that slavery was abolished because our modern ancestors expressed the false belief or the unjustified belief that slavery was wrong. It is not clear in what context the error theorist and moral skeptical explanations would be appropriate answers.

13.8 A New Explanandum

In the previous sections, I criticized Huemer's moral realist explanation of moral progress. However, I acknowledge that he has made a valuable contribution to the debate on moral progress. He has presented an interesting explanandum that metaethicists would be interested in explaining. In his spirit, I present another explanandum that cries out for metaethical explanations.

There are wrong practices around the world today. For example, some countries operate concentration camps where political prisoners are cruelly treated. Some countries develop military robots. Some countries have nuclear weapons. Some countries invade other countries. Presumably, some individuals have recognized that these practices are wrong, but the practices persist. Why do they persist? How do they differ from slavery?

One might answer that non-moral factors, such as the desire for power and the mutual distrust between rivaling countries, are so strong that moral factors do not have enough power to put an end to the current practices. This explanation raises an interesting issue of how strong moral and non-moral factors are in relation to each other, and a related issue of under what conditions moral factors can defeat non-moral factors. Both metaethicists and non-metaethicists might be interested in pursuing these issues.

Moral realists might hypothesize that the property of wrongness in slavery was strong while the property of wrongness in the current practices is weak. This possible hypothesis raises an interesting issue of whether the property of wrongness comes in degrees, and a related issue of how we can discern the different degrees of wrongness. Is the degree of wrongness proportional to the amount of (potential) pain? I leave the task of exploring these issues to those who are sympathetic to the hypothesis that the property of wrongness in slavery was strong while the property of wrongness in the current practices is weak.

How might emotivists, prescriptivists, and quasi-realists answer the foregoing question about the differences between slavery and the current practices? They might suggest that noncognitive states, such as emotions and commands, come in degrees, and then suggest that past moral evaluators expressed *strong* negative emotions and commands toward slavery while current moral evaluators do not express strong negative emotions and commands toward the current practices. They might add that a *large* portion of the general public expressed negative emotions and commands toward slavery while only a small portion of the general public express negative emotions and commands toward the current practices.

How about error theorists and moral skeptics? They might suggest, respectively, that past moral evaluators *strongly* expressed the false belief and the unjustified belief that slavery was wrong while current moral evaluators do not strongly express the false belief and the unjustified belief that the current practices are wrong. They might add, respectively, that a *large* portion of the general public expressed the false belief and the unjustified belief about slavery while only a small portion of the general public express the false belief and the unjustified belief about the current practices. However, none of their explanations sounds convincing to me. Again, do false and unjustified beliefs, in general, contribute to amazing events?

How about cultural emotivists? They might suggest that past moral evaluators expressed the belief that their culture had a *strong* negative emotion toward slavery while current moral evaluators do not express the belief that their culture has strong negative emotions toward the current practices. They might add that a *large* portion of the general public expressed the belief that their culture had a negative emotion toward slavery while only a small portion of the general public express the belief that their culture has negative emotions toward the current practices.

How about moral functionalists? They might suggest that past moral evaluators strongly discouraged slavery while current moral evaluators do not strongly discourage the current practices. They might add that a *large* portion of the general public discouraged slavery while only a small portion of the general public discourage the current practices.

Admittedly, all these metaethical explanations are underdeveloped. My only intention with their introduction is to give future metaethicists an idea how they can explain why some wrong practices perished in the past while some wrong practices persist in the present.

13.9 Conclusion

If moral facts are causally inert, IBE cannot establish their existence. To go further, they cannot serve as explanantia for anything. The moral realist and antirealist explanations of moral progress are metaethical ones, while the sci-tech and reproductive explanations of moral progress are non-metaethical ones. As such, the former can satisfy those who are interested in whether moral factors have contributed to moral progress, while the latter can satisfy those who are interested in whether non-moral factors contributed to moral progress. Finally, an interesting task ahead for future metaethicists is to explain why some wrong practices perished in the past while some wrong practices persist in the present. This chapter can be wrapped up with the following two simple messages to moral realists and antirealists: "No causality, no reality!" and "Invoke moral explanantia!"

References

Baker, A. (2005). Are there genuine mathematical explanations of physical phenomena? *Mind, 114*(454), 223–237.

Baker, A. (2009). Mathematical explanation in science. *The British Journal for the Philosophy of Science, 60*(3), 611–633.

Blackburn, S. (1993). *Essays in quasi-realism*. Oxford University Press.

Chagnon, N. (1988). Life histories, blood revenge, and warfare in a tribal population. *Science, 239*(4843), 985–992.

Friedman, M. (1974). Explanation and scientific understanding. *The Journal of Philosophy, 71*, 5–19.

Hare, R. (1952). *Language of morals*. Clarendon Press.

Hempel, C. (1966). *Philosophy of natural science*. Prentice-Hall.

Huemer, M. (2016). A liberal realist answer to debunking skeptics: The empirical case for realism. *Philosophical Studies, 173*(7), 1983–2010.

Kitcher, P. (1981). Explanatory unification. *Philosophy of Science, 48*(4), 507–531.

Mackie, J. (1977). *Ethics: Inventing right and wrong*. Penguin Books.

Park, S. (2014). Cultural relativism and the theory of relativity. *Filosofija. Sociologija, 25*(1), 44–51.

Rusch, H. (2014). The evolutionary interplay of intergroup conflict and altruism in humans: A review of parochial altruism theory and prospects for its extension. *Proceedings of the Royal Society of London B: Biological Sciences, 281*(1794), 1–9.

Salmon, W. (1978). Why ask, "Why?"? An inquiry concerning scientific explanation. In J. Kourany (Ed.), *Scientific knowledge: Basic issues in the philosophy of science*. Wadsworth Publishing Company.

Schick, T., & Vaughn, L. (2010). *Doing philosophy: An introduction through thought experiments* (4th ed.). Mayfield Publishing Company.

Stevenson, C. (1944). *Ethics and language*. Yale University Press.

Street, S. (2006). A Darwinian dilemma for realist theories of value. *Philosophical Studies, 127*(1), 109–166.

van Fraassen, B. (1980). *The scientific image*. Oxford University Press.

Chapter 14
The Implications of Metaethical Theories for Scientific and Moral Progress

Abstract The five accounts of scientific progress are not only composed of normative assumptions but also of descriptive assumptions. Accordingly, error theory implies that the normative assumptions are false, but not that the descriptive assumptions are false. The same holds for the five accounts of moral progress. Consequently, even if error theory is true, debaters of scientific and moral progress can continue to debate over what is involved in scientific and moral progress. Moreover, there are several alternatives to error theory, and error theory has several problems.

Keywords Moral antirealism · Moral progress · Moral realism · Normative · Scientific progress

14.1 Introduction

Is there progress in science? It appears at least initially that there is. For example, the replacement of the geocentric system with the heliocentric one constitutes scientific progress. But why is it scientific progress? To go further, is scientific progress real? There are rival answers to these questions in the literature on scientific progress. Rival philosophers of scientific progress provide rival accounts of why some scientific changes are progressive; some meta-normative theorists take scientific progress to be real while other meta-normative theorists take it to be unreal.

We can ask similar questions about moral progress. Is there progress in morality? It also appears at least initially that there is. For example, the abolition of slavery constitutes moral progress. But why is it moral progress? To go further, is moral progress real? Rival philosophers of moral progress can provide rival accounts of why some moral changes are progressive; some metaethicists can take moral progress to be real while other metaethicists can take it to be unreal, as I will show in this chapter.

This chapter explores the implications of the eight rival metaethical theories for the five accounts of scientific progress in the literature and for the five accounts of moral progress that I develop in this chapter. The eight metaethical theories are rival

S. Park, *Induction, Science, and Morality*, Synthese Library 520,
https://doi.org/10.1007/978-3-032-00032-3_14

views of what a moral judgment is, whether moral facts exist, and whether we know that some moral statements correspond to moral facts. The five accounts of scientific progress are rival views of what scientific progress consists in. The five accounts of moral progress are rival views of what moral progress consists in. The eight metaethical theories have competing implications for the five accounts of scientific progress and for the five accounts of moral progress.

The present chapter has the following structure. In Sect. 14.2, I explicate the five accounts of scientific progress and then elucidate the implications of the eight metaethical theories for them. In Sect. 14.3, I unpack the five accounts of moral progress and then elucidate the implications of the eight metaethical theories for them. In Sect. 14.4, I expose an important difference between scientific and moral progress. In Sect. 14.5, I explore whether cultural emotivism (Sect. 10.3 of Chap. 10) can help to establish the existence of moral truths and moral progress.

The present chapter is intended to be useful to philosophers who aim to investigate how the eight metaethical theories relate to the five accounts of scientific progress and to the five accounts of moral progress, and to philosophers who aim to develop new accounts of scientific progress, new accounts of moral progress, and new accounts of moral judgment (i.e., new metaethical theories). Readers will see that I criticize moral realism and error theory (Mackie, 1977) in this chapter as I did in previous chapters.

This chapter is inspired by Darrell Rowbottom (2023, Part 3), who takes error theory as implying that there are no objective standards for determining whether some scientific changes are progressive or regressive. It appears that if error theory were true, none of the five rival accounts of scientific progress and none of the five accounts of moral progress would be true. I argue that on close examination, the negative implications of error theory for the five accounts of scientific progress and for the five accounts of moral progress are disputable, and that even if we grant for the sake of argument that they are not disputable, they are limited.

14.2 The Five Accounts of Scientific Progress

14.2.1 The Contents

In this subsection, I sketch five accounts of scientific progress which invoke different explanatory factors to explain why some scientific changes are progressive and why other scientific changes are regressive.

There are many changes in science. Some of them are progressive while others are regressive. For example, the replacement of the geocentric system with the heliocentric one constitutes scientific progress, while ancient Greek philosophers' rejection of atomism constitutes scientific regress. Why does the replacement of the geocentric system with the heliocentric one constitute scientific progress? Why does the rejection of atomism constitute scientific regress?

Five rival answers to this question can be inferred from the five accounts of scientific progress: the problem-solving account (Kuhn, 1962/1970; Laudan, 1977, 1984), the semantic account (Niiniluoto, 1984, 2014), the epistemic account (Bird, 2007), the noetic account (Dellsén, 2016, p. 73), and the evidential account (Park, 2022, pp. 163–166). They make rival claims about what constitutes scientific progress. They might agree or disagree about whether some scientific changes are progressive or regressive. Even if they agree that a particular scientific change is progressive, they provide different explanatory factors to explain why it is progressive. Let me specify them one by one.

The problem-solving account holds that a scientific change is progressive when it involves an increase in problem-solutions. The replacement of the geocentric system with the heliocentric one constituted scientific progress because the heliocentric system solved more problems than did the geocentric system. For example, the heliocentric system solved the problem of planets while the geocentric system could not. The problem of planets is to explain the retrograde motion of planets: planets slow down, stop, move backward, stop, and then move forward in the night sky. On the problem-solving account, a better capacity to solve problems does not mean being closer to the truth.

The semantic account holds that a scientific change is progressive when it involves an increase in verisimilitude. The replacement of the geocentric system with the heliocentric one constituted scientific progress because the heliocentric system was closer to the truth than was the geocentric system. Strictly speaking, the heliocentric system was false, given that it claimed that planets moved around the sun in circles rather than in ellipses. However, the heliocentric system was closer to the truth than the geocentric system, given that the former more correctly represented the structure of the universe than did the latter in other respects.

The epistemic account holds that a scientific change is progressive when it involves an increase in knowledge. The replacement of the geocentric system with the heliocentric one constituted scientific progress because scientists of the heliocentric system knew that the heliocentric system was closer to the truth than was the geocentric system (Bird, 2007, p. 105) and because they knew that the geocentric system was false (Bird, 2007, p. 108). Moreover, the scope of the heliocentric system was wider than that of the geocentric system, and thus the heliocentric system produced more observational knowledge than did the geocentric system.

The noetic account holds that a scientific change is progressive when it involves an increase in understanding. We understand the world more than we did when we explain and predict phenomena more correctly. For example, the heliocentric system explained and predicted phenomena more correctly than did Ptolemaic theory. Hence, the replacement of the geocentric system with the heliocentric one resulted in an increase in understanding. An increase in understanding means getting closer to the truth (Dellsén, 2022, p. 822).

The evidential account holds that a scientific change is progressive when it involves an increase "in evidence for our beliefs about the world" (Park, 2022, p. 163). The replacement of the geocentric system with the heliocentric one constituted scientific progress because scientists had more evidence for the heliocentric system than for the geocentric system. The evidence for the heliocentric system came not only from our observations of the world but also from the refutation of the geocentric system. On the evidential account, the overthrow of the geocentric system counted as evidence for the truth of the heliocentric system. Since scientists had more evidence for the heliocentric system than for the geocentric system, the heliocentric system was closer to the truth (and/or was more likely to be true) than was the geocentric system.

Bear in mind that the five accounts of scientific progress invoke different explanatory factors: increases in problem-solutions, verisimilitude, knowledge, understanding, and evidence.

14.2.2 The Normative Assumptions of the Five Accounts

In this subsection, I elucidate the implications of the eight metaethical theories for the normative assumptions of the five accounts of scientific progress.

The five accounts of scientific progress assume that some scientific changes are progressive while other scientific changes are regressive. What is the nature of these assumptions? They are not descriptive but normative, according to Ilkka Niiniluoto, who writes that "'Progress' is an axiological or a normative concept" (Niiniluoto, 2014, p. 75). Why fuss about the normativity of the assumptions? The eight metaethical theories have competing implications for them. Let me unfold them one by one.

Moral realism (Sect. 10.2 of Chap. 10) holds that we know that some moral statements are true. Moral realists would say that the normative assumptions are true or false. If they endorsed the evidential account, they would say that some scientific changes have the property of goodness because they involve increases in evidence, and that other scientific changes have the property of badness because they involve decreases in evidence. If moral realists endorsed the epistemic account, they would say that some scientific changes have the property of goodness because they involve increases in knowledge, and that other scientific changes have the property of badness because they involve decreases in knowledge.

Emotivism (Stevenson, 1944) asserts that a moral judgment is a moral emotion. Emotivists would say that the positive normative assumption that some scientific changes are progressive is a positive emotion toward those scientific changes, and that the negative normative assumption that other scientific changes are regressive is a negative emotion toward those scientific changes. If emotivists endorsed the evidential account, they would say that philosophers of scientific progress have a positive emotion toward some scientific changes because those scientific changes involve increases in evidence, and that philosophers of scientific progress have a negative emotion toward other scientific changes because those scientific changes involve decreases in evidence.

Prescriptivism (Hare, 1952) is the view that a moral judgment is a moral command. Prescriptivists would say that the positive normative assumption that some scientific changes are progressive is a positive command about those scientific changes, and that the negative normative assumption that other scientific changes are regressive is a negative command about those scientific changes. If prescriptivists endorsed the evidential account, they would say that philosophers of scientific progress issue a positive command about some scientific changes because those scientific changes involve increases in evidence, and that philosophers of scientific progress issue a negative command about other scientific changes because those scientific changes involve decreases in evidence.

Error theory (Mackie, 1977) maintains that there are no moral facts, and thus that no moral judgment is true. Error theorists would say that the positive and negative normative assumptions are all false, given that there are no normative facts that would make them true. As a consequence, the five accounts of scientific progress, which have the positive and negative assumptions as their constituents, are also all false. This is exactly what Rowbottom (2023, Part 3) would say. It appears that error theory has a lethal implication for the five accounts of scientific progress.

Quasi-realism (Blackburn, 1993) claims that moral judgments are noncognitive states, but that we can use "true" in moral discourse because saying that it is true that p is the same as saying that p. Quasi-realists would say that the positive normative assumption about some scientific changes is a positive emotion toward those scientific changes, and that the negative normative assumption about other scientific changes is a negative emotion toward those scientific changes. However, they would add that we can say that the positive and negative assumptions are true. The quasi-realist explanation of why philosophers of scientific progress have the positive emotion toward some scientific changes and the negative emotion toward other scientific changes would be the same as the emotivist explanation above.

Moral skepticism (Street, 2006) is the position that no moral judgment is justified. Moral skeptics would say that none of the positive assumption about some scientific changes and the negative assumption about other scientific changes is justified. They might believe that some scientific changes involve increases or decreases in problem-solutions, verisimilitude, knowledge, understanding, and evidence. However, they would be skeptical that scientific changes that involve these increases or decreases are progressive or regressive, respectively, pointing out that the claims that they are progressive or regressive are normative and thus not justified.

Cultural emotivism (Sect. 10.3 of Chap. 10) holds that moral judgments are beliefs about cultural emotions. For example, when moral evaluators, those who make moral judgments, say, "Blood donation is right," they are expressing the belief that their culture has a positive emotion toward blood donation. When moral evaluators say, "Torture is wrong," they are expressing the belief that their culture has a negative emotion toward torture. In contrast to emotivism, cultural emotivism does not assert that a moral judgment is moral evaluators' emotion about an action, but rather that it is their belief about a cultural emotion.

Cultural emotivists would say that when philosophers of scientific progress say that some scientific changes are progressive, they are expressing the belief that their culture has a positive emotion toward those scientific changes, and that when philosophers of scientific progress say that other scientific changes are regressive, they are expressing the belief that their culture has a negative emotion toward those scientific changes. If cultural emotivists endorsed the evidential account, they would say that philosophers of scientific progress believe that their culture has a positive emotion toward some scientific changes because those scientific changes involve increases in evidence, and that philosophers of scientific progress believe that their culture has a negative emotion toward other scientific changes because those scientific changes involve decreases in evidence.

Moral functionalism (Sects. 11.2 and 11.5 of Chap. 11) asserts that moral utterances perform the functions of encouraging and discouraging moral actions, actions that are either right or wrong. For example, the moral utterance "Blood donation is right" performs the function of encouraging blood donation, while the moral utterance "Torture is wrong" performs the function of discouraging torture. Moral functionalism is noncommittal about whether moral judgments are truth-evaluable and about whether moral properties are real.

Moral functionalists would say that the positive normative assumption that some scientific changes are progressive performs the function of encouraging those scientific changes while the negative normative assumption that other scientific changes are regressive performs the function of discouraging those scientific changes. If moral functionalists endorsed the evidential account, they would say that philosophers of scientific progress encourage some scientific changes because those scientific changes involve increases in evidence, and that philosophers of scientific progress discourage other scientific changes because those scientific changes involve decreases in evidence.

So far, I argued that the eight metaethical theories have competing implications for the positive normative assumption that some scientific changes are progressive and for the negative normative assumption that other scientific changes are regressive. From now on, I explore the challenges that the existence of the rival implications poses to moral realism and error theory.

Let me begin with moral realism. Suppose that moral realists construct the following argument for moral realism. Some scientific changes are progressive because they have the property of goodness, and other scientific changes are regressive because they have the property of badness. In other words, it is true that some scientific changes are progressive because it corresponds to a normative fact that those scientific changes are good, and it is true that other scientific changes are regressive because it corresponds to a normative fact that those scientific changes are bad. Therefore, the existence of scientific progress and regress shows that there are normative properties. In short, scientific progress and regress do not make sense unless normative properties inhere in some scientific changes.

How would moral antirealists respond to the foregoing moral realist suggestion that some scientific changes are progressive because they have the property of goodness, and that other scientific changes are regressive because they have the property of badness? Emotivists, for example, would say that there are no normative properties, and that moral realists are expressing their positive emotion toward some scientific changes and their negative emotion toward other scientific changes. Accordingly, moral realists have the burden of showing that their response to the normative assumptions is more convincing than other metaethicists' responses to them.

Let me turn to error theory. Rowbottom endorses the negative implication of error theory for the five accounts of scientific progress, writing that "There are no objective standards (or criteria) that determine which changes in science are cognitively good or progressive" (Rowbottom, 2023, p. 89). Error theory implies that there are no normative facts that would make the normative assumptions of the five accounts true, and thus that none of the five accounts is true.

Rowbottom has the burden of demonstrating that the implication of error theory for the five accounts is more convincing than those of other metaethical theories for the five accounts. For example, moral functionalism implies that the normative assumptions perform the functions of encouraging some scientific changes and discouraging other scientific changes, but not that the normative assumptions are false. What would be the reason for thinking that the implication of error theory is more convincing than that of moral functionalism?

Alternatively, Rowbottom has the burden of showing that error theory is better than other metaethical theories. This burden is a heavy one, given that I argued in Chaps. 11, 12, and 13, error theory is vulnerable to several objections while other metaethical theories are not. Let me briefly remind readers of what my objections were in those chapters.

In Sect. 11.4.1 of Chap. 11, I argued that error theory is the worst of the eight metaethical theories in terms of effectiveness in preventing moral agents from committing wrong acts. Imagine that religious fanatics resolve to kill heretics, and that error theorists say to them, "Killing heretics is wrong. But my moral belief is false." Uttering this statement would only add fuel to religious fanatics' desire to kill heretics.

In Sect. 12.3 of Chap. 12, I argued that (a) error theory cannot explain why some moral utterances are widely used while other moral sentences are not in our moral lives, that (b) error theory cannot spell out a difference between a moral judgment and its opposite moral judgment, say, between "Blood donation is right" and "Blood donation is wrong," and that (c) error theory violates the principle of charity which implies that we should not ceteris paribus attribute false beliefs to others. These three objections to error theory also apply to moral skepticism, but not to other metaethical theories.

In Sect. 13.6 of Chap. 13, I argued that error theory cannot provide a metaethical explanation of why morality has progressed. It does not sound convincing for error theorists to suggest, for example, that slavery was abolished because our modern ancestors expressed the false belief that slavery was wrong. After all, it is questionable whether the expression of false beliefs, in general, contributes to the occurrence of amazing events, such as the abolition of slavery.

More problems with error theory will be exposed in Chap. 15, where I criticize the evolutionary debunking argument for moral skepticism (Street, 2006). I object, for example, that both error theory and moral skepticism can lead to paradoxes, and that they are saddled with the burden of explaining why we have the maladaptive traits: the capacity to form false beliefs and the capacity to form unjustified beliefs.

To summarize this subsection, the negative implication of error theory for the normative assumptions of the five accounts of scientific progress is disputable, given that it competes with those of other metaethical theories.

14.2.3 The Descriptive Assumptions of the Five Accounts

In this subsection, I argue that error theory does not imply that the descriptive assumptions of the five accounts of scientific progress are false.

Each of the five accounts of scientific progress is composed not only of normative assumptions but also descriptive assumptions. For example, the problem-solving account holds that some scientific changes are progressive because they involve increases in problem-solutions, and that other scientific changes are regressive because they involve decreases in problem-solutions. The two explananda are normative assumptions while the two explanantia are descriptive assumptions. The same holds for other accounts of scientific progress.

Error theory implies that the explananda of the five accounts, which are normative assumptions, are false, but not that the explanantia of the five accounts, which are descriptive assumptions, are false. Specifically, error theory implies that it is false that scientific changes are progressive or regressive, but not that they involve increases or decreases in problem-solutions, verisimilitude, knowledge, understanding, or evidence. Therefore, error theory does not imply that the five accounts are completely false, and it is an open question whether they are approximately true.

This conclusion has the following interesting implication. Even if error theory is true, debaters of scientific progress do not need to despair and stop debating about what is involved in scientific changes. They can keep debating about whether some scientific changes involve increases in problem-solutions, verisimilitude, knowledge, understanding, or evidence. They can even develop new accounts of scientific progress which will invoke new explanatory factors. In this sense, the negative implication of error theory for the five accounts of scientific progress about scientific progress is limited.

14.3 The Five Accounts of Moral Progress

14.3.1 The Contents

In this subsection, I sketch five accounts of moral progress which invoke different explanatory factors to explain why some moral changes are progressive or regressive.

Some moral changes are progressive while other moral changes are regressive. For example, slavery was practiced in ancient times, but was abolished in modern times. The abolition of slavery constituted moral progress. In contrast, freedom of speech might be restricted in a country with the inauguration of a dictatorial president today. The restriction of free speech constitutes moral regress. Just as philosophers of science develop rival accounts of scientific progress, so metaethicists can develop rival accounts of moral progress. Let me unfold them one by one.

Louis Pojman writes, "If we can be closer to the truth regarding factual or scientific matters, why can't we be closer to the truth on moral matters?" (Pojman, 2008, p. 21). In Pojman's spirit, some philosophers of moral progress might suggest that moral progress consists in an increase in verisimilitude, and that moral regress consists in a decrease in verisimilitude. This suggestion can be called "the moral semantic account." On the moral semantic account, the abolition of slavery constituted moral progress because we got closer to the moral truth that slavery was wrong, while the restriction of free speech constitutes moral regress because we get farther from the moral truth that free speech is right.

Some philosophers of moral progress might suggest that moral progress consists in an increase in happiness, and that moral regress consists in a decrease in happiness. This suggestion can be called "the utilitarian account." On the utilitarian account, the abolition of slavery constituted moral progress because it involved an increase in happiness, while the restriction of free speech constitutes moral regress because it involves a decrease in happiness.

Some philosophers of moral progress might suggest that moral progress consists in an increase in justice, and that moral regress consists in a decrease in justice. This suggestion can be called "the justice account." On the justice account, the abolition of slavery constituted moral progress because it involved an increase in justice, while the restriction of free speech constitutes moral regress because it involves a decrease in justice.

Some philosophers of moral progress might suggest that moral progress consists in an increase in freedom, and that moral regress consists in a decrease in freedom. This suggestion can be called "the freedom account." On the freedom account, the abolition of slavery constituted moral progress because it involved an increase in freedom, while the restriction of free speech constitutes moral regress because it involves a decrease in freedom.

Some philosophers of moral progress might suggest that moral progress consists in an increase in equality, and that moral regress consists in a decrease in equality. This suggestion can be called "the equality account." On the equality account, the abolition of slavery constituted moral progress because it involved an increase in equality, while the restriction of free speech constitutes moral regress because it involves a decrease in equality.

There can be a similarity between the five accounts of moral progress and the five accounts of scientific progress. As noted in Sect. 14.2.1 above, the exponents of the five accounts of scientific progress might agree or disagree about whether some moral changes are progressive or regressive. Analogously, the proponents of the five accounts of moral progress might agree or disagree about whether some moral changes are progressive or regressive.

There can be a further similarity between the five accounts of moral progress and the five accounts of scientific progress. Even if the exponents of the five accounts of scientific progress agree about whether a particular scientific change is progressive or regressive, they disagree about why it is progressive or regressive, providing different explanatory factors to explain why it is progressive or regressive. Similarly, even if the proponents of the five accounts of moral progress agree about whether a particular moral change is progressive or regressive, they can disagree about why it is progressive or regressive, providing different explanatory factors to explain why it is progressive or regressive.

Admittedly, the five accounts of moral progress sketched in this subsection are underdeveloped. I leave to future researchers the task of transforming them into full-fledged accounts of moral progress, as well as the task of unearthing episodes that support one account of moral progress but undermine other accounts of moral progress. Accomplishing these tasks will provide us with an insight into the nature of moral progress.

14.3.2 The Normative Assumptions of the Five Accounts

In this subsection, I clarify the implications of the eight metaethical theories for the positive normative assumption that some moral changes are progressive and for the negative normative assumption that other moral changes are regressive.

What are philosophers of moral progress doing when they say that some moral changes are progressive while other moral changes are regressive? They are not expressing descriptive assumptions but normative assumptions. As Toni Rønnow-Rasmussen writes, "claims to the effect that someone or something has (or has not) made moral progress are evaluative terms about certain kinds of changes" (Rønnow-Rasmussen, 2017, p. 137). Consequently, the eight metaethical theories have competing implications for the positive and negative normative assumptions.

The competing implications can be briefly put as follows. Moral realism implies that the positive and negative normative assumptions are true. Emotivism implies that they are positive and negative emotions. Prescriptivism implies that they are positive and negative commands. Error theory implies that none of them is true. Quasi-realism implies that they are positive and negative emotions, but we can say that they are true. Moral skepticism implies that none of them is justified. Cultural emotivism implies that they are beliefs about positive and negative cultural emotions. Moral functionalism implies that they perform the functions of encouraging some moral changes and discouraging other moral changes.

Metaethicists espousing the eight rival metaethical theories might accept different accounts of moral progress. As a result, they might respond differently to the positive and negative normative assumptions. For example, cultural emotivists, who have accepted the utilitarian account of moral process, would say that philosophers of moral progress express the belief that their culture has a positive emotion toward some moral changes because those moral changes involve increases in happiness, and the belief that their culture has a negative emotion toward other moral changes because those moral changes involve decreases in happiness. Moral functionalists, who have accepted the justice account of moral process, would say that philosophers of moral progress encourage some moral changes because those moral changes involve increases in justice, and discourage other moral changes because those moral changes involve decreases in justice.

So far, I argued that the eight metaethical theories have competing implications for the positive normative assumption that some moral changes are progressive and for the negative normative assumption that other moral changes are regressive. From now on, I explore the challenges that the existence of the rival implications poses to moral realism and error theory.

Let me begin with moral realism. Moral realists might construct the following argument for moral realism. Some moral changes are progressive while other moral changes are regressive. If no moral properties were real, there could neither be moral progress nor moral regress. After all, some moral changes are progressive because they have the property of rightness, and other moral changes are regressive because they have the property of wrongness. Therefore, they have moral properties. In short, moral progress and regress do not make sense unless there are moral properties.

Moral antirealists, such as cultural emotivists, would object that moral realists are expressing the belief that their culture has a positive emotion toward some moral changes when they say that those moral changes are progressive, and that moral realists are expressing the belief that their culture has a negative emotion toward other moral changes when they say that those moral changes are regressive. Consequently, moral realists have the burden of showing that their response to the normative assumptions is more convincing than those of other metaethicists' responses to them.

Let me turn to error theory. Error theorists might construct the following argument against the five accounts of moral progress. The five accounts of moral progress have, as their constituents, the positive normative assumption that some moral changes are progressive, and the negative normative assumption that other moral changes are regressive. Given that no normative assumption is true, none of the five accounts of moral progress is true.

Other metaethicists would object that the implication of error theory competes with those of other metaethical theories. For example, it competes with that of moral functionalism according to which the normative assumptions perform the functions of encouraging some moral changes and discouraging other moral changes. Therefore, those who endorse the implication of error theory have the burden of showing that it is more convincing than are those of other metaethical theories.

To sum up this subsection, the negative implication of error theory for the normative assumptions of the five accounts of moral progress is disputable, given that it competes with those of other metaethical theories.

14.3.3 The Descriptive Assumptions of the Five Accounts

In this subsection, I argue that error theory does not imply that the descriptive assumptions of the five accounts of moral progress are false.

Each of the five accounts is composed not only of normative assumptions but also of descriptive assumptions. For example, the utilitarian account holds that some moral changes are progressive because they involve increases in happiness, and that other moral changes are regressive because they involve decreases in happiness. The two explananda are normative assumptions, but the two explanantia are descriptive assumptions. The same holds for other accounts of moral progress.

Error theory implies that the normative assumptions are false, but not that the descriptive assumptions are false. Therefore, even if we grant for the sake of argument that the error theory is true, it is an open question whether the five accounts of moral progress are approximately true, and philosophers of moral progress can debate about whether some moral changes involve increases or decreases in verisimilitude, happiness, justice, freedom, equality, or something else. In this sense, the negative implication of error theory for the five accounts is limited.

14.4 A Similarity Between the Two Debates

Some writers claim that just as we can get closer to scientific truths, so we can get closer to moral truths. In this section, I point out a similarity between the debate about scientific progress and the debate about moral progress. The similarity will present a challenge to those who compare moral progress and scientific progress for their theoretical purposes.

Do we get closer to scientific truths when some scientific changes occur? The answer is "Yes," according to the semantic, epistemic, noetic, and evidential accounts. They claim that some scientific changes involve increases in verisimilitude, knowledge, understanding, and evidence, respectively. An increase in verisimilitude means getting closer to truth. An increase in knowledge entails the arrival at truth. An increase in understanding indicates getting closer to truth. An increase in evidence indicates getting closer to truth and/or being more likely to be true.

How about the problem-solving account? As noted in Sect. 14.4 above, it denies that when scientific progress occurs, we move closer to scientific truths. For example, the replacement of the geocentric system with the heliocentric one constituted scientific progress, but that does not mean that heliocentric scientists were closer to the scientific truth than were geocentric scientists; it only means that heliocentric scientists solved more problems than did geocentric scientists.

As just noted, it is controversial whether scientific progress involves getting closer to scientific truths. In consequence, it is problematic for moral realists to appeal to scientific progress to establish that morality progresses toward moral truths. The espousers of the problem-solving account would reject moral realists' strategy, saying that science does not move toward scientific truths, so it is problematic to say that just as science moves toward scientific truths, so morality moves toward moral truths.

How about morality? Does morality move toward moral truths? My answer is that just as it is controversial among philosophers of scientific progress whether science moves toward scientific truths, so it would be controversial among philosophers of moral progress whether morality moves toward moral truths. This answer is fleshed out as follows.

On the moral semantic account, some moral changes involve getting closer to moral truths. Moral realists would accept the moral semantic account while moral antirealists would reject it. For example, error theorists would argue that since no moral fact exists, no moral truth exists, and that we can never get closer to moral truths no matter how right moral changes might be. Moral skeptics would say that even if moral facts exist, we cannot know that they do, and that we cannot know that we get closer to moral truths no matter how right moral changes might be. In short, error theorists deny, and moral skeptics are skeptical, that some moral changes involve getting closer to moral truths.

The utilitarian, justice, freedom, and equality accounts of moral progress assert that moral progress consists in increases in happiness, justice, freedom, and equality, respectively. None of them claims that moral progress involves getting closer to moral truths. They are similar to the exponents of the problem-solving account of scientific progress, who deny that scientific progress involves getting closer to scientific truths.

In sum, there is a similarity between the debate about scientific progress and the debate about moral progress. Some philosophers of scientific progress affirm, while other philosophers of scientific progress deny, that scientific progress involves getting closer to scientific truths. Similarly, some philosophers of moral progress affirm, while other philosophers of moral progress deny, that moral progress involves getting closer to moral truths.

This similarity between the two debates presents the following challenges to moral realists and error theorists. Given that it is controversial whether science moves toward scientific truths, it is problematic for moral realists to say that just as science moves toward scientific truths, so morality moves toward moral truths. In addition, given that it is also controversial whether morality does not move toward moral truths, it is also problematic for error theorists to say that science does not move toward scientific truths any more than morality moves toward moral truths.

14.5 Fiddling with the Definition of "Moral Fact"

Some metaethicists might define "moral fact" as cultural emotion, and then suggest that there are moral truths, and that a moral change is progressive or regressive, depending on whether it involves getting closer to or farther from a moral truth. I block this possible move in this section.

Cultural emotivism holds that a moral judgment is a belief about a cultural emotion. It is true or false, depending on whether it corresponds to a cultural emotion. For example, it is true that charitable giving is right because it corresponds to a positive cultural emotion toward charitable giving, and it is false that charitable giving is wrong because it does not correspond to a positive cultural emotion toward charitable giving. Cultural emotivism affirms that we know that some moral beliefs are true, but denies that moral facts, which are putatively mind-independent parts of the world, make moral beliefs true or false.

Those who wish to believe that moral facts exist might define "moral fact" as cultural emotion, and then claim, as do cultural emotivists, that a moral judgment is a belief about a cultural emotion. The new definition and cultural emotivism jointly yield the proposal that a moral judgment is true if and only if it corresponds to a moral fact, i.e., to a cultural emotion. For example, it is true that charitable giving is right because it corresponds to a moral fact; it is false that charitable giving is wrong because it does not correspond to a moral fact.

Let me call this proposal "definitional cultural emotivism." It is so named because it attempts to establish the existence of moral truths by fiddling with the definition of "moral fact." Definitional cultural emotivists newly define "moral fact" so that they can achieve their theoretical purpose of establishing the existence of moral truths. After establishing the existence of moral truths in their own manner, they suggest that a moral change is progressive or regressive, depending on whether it involves getting closer to or farther from a moral truth.

What are we to make of definitional cultural emotivism? My response to it is to argue that we should be careful when we interpret definitional cultural emotivists' metaethical utterances. They say, for example, "Some moral beliefs are made true by moral facts." This utterance does not mean that some moral beliefs are made true by mind-independent parts of the world; it rather means that some moral beliefs are made true by cultural emotions. Their metaethical utterance "Some moral facts exist, and morality progresses toward moral truths" is not committed to the existence of mind-independent parts of the world that would make some moral beliefs true; it is rather committed to the existence of cultural emotions that make some moral beliefs true.

Relatedly, definitional cultural emotivists cannot say that morality progresses toward moral truths in the way science progresses toward scientific truths. When philosophers of scientific progress say that science progresses toward scientific truths, they mean that science gets closer to scientific beliefs that are made true by mind-independent parts of the world. However, when definitional cultural emotivists say that morality progresses toward moral truths, they do not mean that morality gets closer to moral beliefs that are made true by mind-independent parts of the world, but rather that morality gets closer to moral beliefs that are made true by cultural emotions. We should guard against the slide from cultural emotions to mind-independent parts of the world.

I ask a simple question to metaethicists who aim to establish that we have moral truths by fiddling with the definition of "moral fact." Why not define "moral fact" as moral belief? This simple definition provides them with a quick way to establish that we have moral truths, and that moral progress consists in an increase in verisimilitude. They would retort that it is unreasonable to define "moral fat" as moral belief. However, this retort invites the following question. Why is it unreasonable to define "moral fact" as moral belief, but reasonable to define it as cultural emotion?

14.6 Conclusion

The eight rival metaethical theories have rival implications for the positive normative assumption that some scientific changes are progressive and for the negative normative assumption that other scientific changes are regressive. They also have rival implications for the positive normative assumption that some moral changes are progressive and for the negative normative assumption that other moral changes are regressive.

The five accounts of scientific progress and the five accounts of moral progress have the positive and negative assumptions as their constituents. In consequence, the eight rival metaethical theories have rival implications for the five accounts of scientific progress and for the five accounts of moral progress. However, they also have descriptive assumptions as their constituents for which the eight rival metaethical theories have no implication.

This chapter has the following key messages to moral realists and error theorists: "Neither scientific progress nor moral progress helps moral realism," and "Error theory doesn't terminate the debates over scientific and moral progress."

References

Bird, A. (2007). What is scientific progress? *Noûs, 41*(1), 92–117.

Blackburn, S. (1993). *Essays in quasi-realism*. Oxford University Press.

Dellsén, F. (2016). Scientific progress: Knowledge vs. understanding. *Studies in History and Philosophy of Science, 56*, 72–83.

Dellsén, F. (2022). Thinking about progress: From science to philosophy. *Noûs, 56*(4), 814–840.

Hare, R. (1952). *Language of morals*. Clarendon Press.

Kuhn, T. (1962/1970). *The structure of scientific revolutions*. University of Chicago Press.

Laudan, L. (1977). *Progress and its problems: Towards a theory of scientific growth*. University of California Press.

Laudan, L. (1984). *Science and values*. University of California Press.

Mackie, J. (1977). *Ethics: Inventing right and wrong*. Penguin Books.

Niiniluoto, I. (1984). *Is science progressive?* Reidel.

Niiniluoto, I. (2014). Scientific progress as increasing verisimilitude. *Studies in History and Philosophy of Science, 46*, 73–77.

Park, S. (2022). *Embracing scientific realism*. Springer.

Pojman, L. (2008). Who's to judge what's right or wrong? In S. Satris (Ed.), *Taking sides: Clashing views on moral issues* (pp. 13–21). McGraw-Hill Companies.

Rønnow-Rasmussen, T. (2017). On locating value in making moral progress. *Ethical Theory and Moral Practice, 20*(1), 137–152.

Rowbottom, D. (2023). *Scientific progress*. Cambridge University Press.
Stevenson, C. (1944). *Ethics and language*. Yale University Press.
Street, S. (2006). A Darwinian dilemma for realist theories of value. *Philosophical Studies, 127*(1), 109–166.

Chapter 15
The Evolutionary Debunking Argument for Moral Skepticism

Abstract Moral skeptics argue that their evolutionary explanation is more parsimonious than moral realists' evolutionary explanation of why we have the mental capacity to make moral judgments, and that their evolutionary explanation supports moral skepticism, according to which no moral judgment is justified. I object that there are several alternative metaethical views whose evolutionary explanations are as economical as moral skeptics' evolutionary explanation. In addition, I offer six objections to moral skepticism and error theory, and then argue that none of them applies to several alternative metaethical views.

Keywords Error theory · Skeptic's paradox · Evolutionary theory · Justified belief · Moral skepticism

15.1 Introduction

Moral skepticism holds that no moral judgment is justified, while moral realism (Sect. 10.2 of Chap. 10) holds that we know that some moral judgments are true. Moral skeptics advance an evolutionary[1] debunking argument to undermine moral realism and to support moral skepticism. The debunking argument holds that we have the mental capacity to make moral judgments not because it tracked moral facts but rather because it enhanced fitness, and thus that no moral judgment is justified.

This chapter aims to show that there are several alternative metaethical views which are preferable to moral skepticism and error theory. In Sect. 15.2, I compare moral skeptics' evolutionary explanation and moral realists' evolutionary explanation of why we have the capacity to make moral judgments. In Sect. 15.3, I argue that rival metaethicists have rival views about what needs to be explained from the evolutionary point of view. In Sect. 15.4, I offer six objections to moral skepticism

[1] I drop "evolutionary" hereafter.

S. Park, *Induction, Science, and Morality*, Synthese Library 520,
https://doi.org/10.1007/978-3-032-00032-3_15

and error theory, and then show that none of those objections applies to other metaethical theories.

This chapter is intended to be beneficial to metaethicists who wonder whether their metaethical theories mesh well with or conflict with evolutionary theory, and whether their metaethical theories are preferable to moral skepticism and error theory.

15.2 The Two Explanations

In this section, I unpack moral skeptics' evolutionary explanation and moral realists' evolutionary explanation of why we have the capacity to make moral judgments, and then introduce an argument for the preference of moral skeptics' evolutionary explanation to moral realists' evolutionary explanation.

We have the capacity to make moral judgments, such as the judgment that donating blood is (morally) right and the judgment that torture is (morally) wrong. Why do we have this capacity? Evolutionary theory provides the following answer. A variation occurred to our ancestors in the distant past. As a result of that variation, some babies were born with the capacity. A group whose members had the capacity did better in the battle to survive and reproduce than a group whose members did not have the capacity. The capacity was passed on to next generations. After many generations, those with the capacity persisted while those without the capacity perished. In short, we have the capacity because it was fitness-conducive.

Both moral realists and skeptics endorse the foregoing evolutionary explanation of why we have the capacity to make moral judgments. They both believe that the capacity increased our ancestors' viability and fertility. However, they diverge on the issue of whether to explain why the capacity increased our ancestors' viability and fertility. Moral realists suggest, while moral skeptics do not, that the capacity increased our ancestors' viability and fertility because it tracked moral facts. Keep in mind that moral realists provide, while moral skeptics do not, the additional explanation that the capacity was an adaptive trait because it tracked moral facts.

On what grounds do moral skeptics reject moral realists' evolutionary explanation that we have the capacity to make moral judgments ultimately because it tracked moral facts? They argue that their evolutionary explanation is more parsimonious than moral realists' evolutionary explanation (Street, 2006, p. 129). The idea is that we need the hypothesis that the capacity was fitness-enhancing, but not the additional hypothesis that the capacity tracked moral facts, to explain why we have the capacity. Thus, Ockham's Razor enjoins us from invoking moral facts to explain why we have moral beliefs.

Let me make a quick comment in defense of moral realism. Moral realism explains not only why we have the capacity to form moral beliefs but also why it was an adaptive trait. By contrast, moral skepticism explains why we have the capacity to form moral beliefs, but not why it was an adaptive trait. Admittedly, moral realism is more complex than is moral skepticism in that moral realism claims, while moral skepticism does not, that the capacity tracked moral facts.

However, moral realism explains more phenomena than does moral skepticism. Therefore, it is problematic for debunkers to say that moral skepticism is better than moral realism.

At any rate, we should bear in mind the following two aspects of moral skepticism. First, it is allegedly justified by the evolutionary explanation of why we have the capacity to make moral judgments. Second, it is built upon cognitivism, the view that moral judgments are truth-apt. After all, it asserts that no moral judgment is justified, which presupposes that moral judgments are a kind of mental states that can be justified, i.e., that can be shown to be likely to be true[2] by evidence. Keep these two aspects of moral skepticism in mind, as they are the targets of my criticisms below.

15.3 Diverse Explananda

As we have seen in Sect. 15.2 above, moral realists and skeptics both assume that we need to explain why we have the capacity to form moral beliefs. In this section, I argue that other metaethicists would disagree with moral realists on this count, and that they would provide their own evolutionary explanations of what they take to be the right explananda.

Let me begin with emotivism (Stevenson, 1944). It holds that moral judgments are emotions. For example, the moral judgment that donating blood is right is a positive emotion toward donating blood, and the moral judgment that torture is wrong is a negative emotion toward torture. Moral judgments are not a kind of mental states that can be made true by moral facts. Nor are they a kind of mental states that can be shown to be likely to be true by evidence, i.e., that can be epistemically justified.

Emotivists would claim that what needs to be explained is not the capacity to form moral beliefs but the capacity to have moral emotions, and then suggest that we have the capacity to have moral emotions because it was fitness-conducive. Emotivists would reject the realist explanation that it was fitness-enhancing because it tracked moral facts. In addition, just as moral skeptics claim that moral skepticism is justified by their evolutionary explanation of why we have the capacity to form moral beliefs, so emotivists would claim that emotivism is justified by their evolutionary explanation of why we have the capacity to have moral emotions.

Let me move on to prescriptivism (Hare, 1952). It holds that a moral sentence expresses a command about a moral action. For example, the moral sentences "Donating blood is right" and "Torture is wrong" are equivalent to the imperative sentences "Donate your blood" and "Don't torture," respectively. An imperative sentence is not a kind of sentence that can be made true by a moral fact. Nor is it a kind of sentence that can be shown to be likely to be true by evidence.

[2] Alvin Goldman and Bob Beddor (2021) claim that a justified belief is likely to be true.

Prescriptivists would claim that what needs to be explained is not the capacity to form moral beliefs but the capacity to issue moral commands, and then suggest that we have the capacity to issue moral commands because it was fitness-enhancing. Prescriptivists would reject the realist explanation that it was fitness-enhancing because it tracked moral facts. In addition, they would claim that prescriptivism is justified by their evolutionary explanation of why we have the capacity to issue commands.

Error theory (Mackie, 1977) holds that no moral fact exists, that moral judgments are beliefs, and thus that they are false. For example, it is false that donating blood is right. It is also false that torture is wrong. We are in error about the world when we believe that moral facts inhabit the world. Given that error theorists believe that no moral fact exists, they would with moral skeptics that no moral judgment is justified.

Error theorists would claim that what needs to be explained is the capacity to form false beliefs (moral judgments), and then suggest that we have the capacity to form false beliefs because it was advantageous for our ancestors' survival and reproduction. Error theorists would reject the realist explanation that it was advantageous for our ancestors' survival and reproduction because it tracked moral facts. In addition, they would claim that error theory is justified by their evolutionary explanation of why we have the capacity to form false beliefs.

Let me turn to quasi-realism (Blackburn, 1993). It holds that a moral judgment is a noncognitive state, but that we can say that it is true because saying that p is true is equivalent to saying that p. Moral evaluators, those who make moral judgments, are expressing their positive and negative attitudes toward donating blood, respectively, when they say that donating blood and torture are right and wrong. However, they can say that it is true that donating blood and torture are right and wrong, respectively, because to say so is equivalent to saying that donating blood and torture are right and wrong, respectively.

Quasi-realists would claim that what needs to be explained is not the capacity to form moral beliefs but the capacity to have moral emotions, and then suggest that we have the capacity to have moral emotions because it increased our distant ancestors' viability and fertility. Quasi-realists would reject the realist explanation that it increased our distant ancestors' viability and fertility because it tracked moral facts. In addition, they would claim that quasi-realism is justified by their evolutionary explanation of why we have the capacity to have moral emotions.

Cultural emotivism (Sect. 10.3 of Chap. 10) holds that a moral judgment is a belief about a cultural emotion. The moral judgment that donating blood is right represents the positive cultural emotion toward donating blood, and the moral judgment that torture is wrong represents the negative cultural emotion toward torture. A moral statement is true or false, depending on whether or not it correctly describes a cultural emotion. In addition, a moral statement is justified or unjustified, depending on whether or not evidence is presented for it.

Cultural emotivists would claim that what needs to be explained is our capacity to form beliefs about cultural emotions, and then suggest that we have the capacity because it was an adaptive trait, and that it was an adaptive trait because it tracked cultural emotions. On this additional explanation, those who could read cultural emotions toward actions were more likely to survive and reproduce than those who could not. In addition, cultural emotivists would claim that cultural emotivism is justified by their evolutionary explanation of why we have the capacity to report cultural emotions.

Let me make a favorable comment about cultural emotivism. Like moral realism, cultural emotivism explains not only why we have the capacity to make moral judgments but also why it was an adaptive trait. As a result, the explanatory scope of cultural emotivism is as wide as that of moral realism. In addition, cultural emotivism has an advantage over moral realism. Moral realism invokes moral facts, while cultural emotivism invokes cultural emotions, to explain why the capacity to make moral judgments was fitness-enhancing. Moral facts are metaphysically dubious whereas cultural emotions are not. Therefore, cultural emotivists' evolutionary explanation is more believable than moral realists' one.

Moral functionalism (Sects. 11.2 and 11.5 of Chap. 11) holds that a moral judgment performs the function of encouraging or discouraging moral actions, actions that can be either right or wrong. For example, the moral utterance "Donating blood is right" performs the function of encouraging donating blood, and the moral utterance "Torture is wrong" performs the function of discouraging torture. Moral functionalism is noncommittal about whether moral judgments are cognitive, whether moral facts exist, and whether we have moral knowledge.

Moral functionalists would claim that what needs to be explained is our capacity to encourage and discourage moral actions, and then suggest that we have the capacity because it increased the chance that our distant ancestors survived and reproduced. Moral functionalists would be noncommittal about the additional explanation that it was fitness-enhancing because it tracked moral facts. In addition, they would claim that moral functionalism is justified by their evolutionary explanation of why we have the capacity to influence behaviors.

So far, I have sketched diverse evolutionary explanations of why we have the capacity to make moral judgments. Rival metaethicists holding these explanations have different views about what the capacity to make moral judgments is. Moral realists, skeptics, and error theorists believe that it is the capacity to form moral beliefs. Emotivists and quasi-realists believe that it is the capacity to have moral emotions. Prescriptivists believe that it is the capacity to have moral commands. Cultural emotivists and moral functionalists believe that it is the capacity to form beliefs about cultural emotions and the capacity to regulate moral actions, respectively.

So what? Moral skeptics' evolutionary explanation is not the only game in town, and moral skeptics have the burden of demonstrating that their evolutionary explanation is preferable to other metaethicists' evolutionary explanations. Moral skeptics share with moral realists and error theorists the view that the capacity to make moral judgments about actions is the capacity to form moral beliefs about actions. Noncognitivists would object that moral skeptics conceded too much to moral realists and error theorists.

15.4 Six Objections to Moral Skepticism and Error Theory

15.4.1 Self-Defeating

In this subsection, I argue that it is potentially self-defeating for moral skeptics and error theorists, but not for other metaethicists, to appeal to evolutionary theory to explain why we have the capacity to make moral judgments.

Helen De Cruz et al. (2011, p. 518) offer a conceptual objection to the debunking argument against our cognitive faculties. They observe that it is potentially self-defeating to use evolutionary theory, which is a product of our cognitive faculties, to establish that our cognitive faculties are not reliable. They ask, "if our cognitive faculties are deeply unreliable, why should we buy into evolutionary theory, which is after all a product of those same cognitive faculties?" (De Cruz et al., 2011, p. 518).

De Cruz et al.'s objection to the debunking argument against our cognitive faculties also applies to the debunking argument for moral skepticism. As noted at the end of Sect. 15.2 above, moral skepticism is committed to cognitivism. On cognitivism, moral judgments are truth-evaluable just as are descriptive beliefs, such as the belief that natural selection has occurred. If no moral judgment is justified, no descriptive belief is justified either, and thus it is self-defeating for moral skeptics to appeal to evolutionary theory, a set of unjustified beliefs, to argue that no moral judgment is justified.

Moral skeptics would reply that we have not only the capacity to form unjustified beliefs but also the capacity to form justified beliefs. Moral judgments are products of the capacity to form unjustified beliefs while descriptive beliefs, such as the belief that species have evolved, are products of the capacity to form justified beliefs. Accordingly, no moral judgment is justified while evolutionary theory (a set of descriptive beliefs) is justified, and it is not self-defeating to rely on evolutionary theory to argue that no moral judgment is justified.

However, it is an ad hoc assumption that moral judgments are products of the capacity to form unjustified beliefs while descriptive beliefs are products of the capacity to form justified beliefs. This assumption is made solely for the sake of averting the conceptual objection that it is self-defeating to appeal to evolutionary theory to show that no moral belief is justified. The ad hoc assumption competes with another ad hoc assumption that both moral judgments and descriptive beliefs are products of the capacity to form justified beliefs. It is not clear how moral skeptics could argue their ad hoc assumption is more plausible than the other ad hoc assumption.

As noted in Sect. 15.2 above, moral realists' evolutionary explanation holds that we have the capacity to form moral beliefs because it is an adaptive trait, and that it is an adaptive trait because it tracks moral facts. Unlike moral skeptics' evolutionary explanation, moral realists' evolutionary explanation does not indicate that no descriptive belief is justified; it rather indicates that since some descriptive beliefs (moral beliefs) are justified, other descriptive beliefs, including evolutionary theory, are also justified. In consequence, it is not self-defeating for moral realists to appeal to evolutionary theory to support moral realism.

Emotivists' evolutionary explanation holds that we have the capacity to have moral emotions because it was fitness-enhancing. This explanation concerns the capacity to have moral emotions, but not the capacity to form descriptive beliefs. It does not at all indicate that no descriptive belief is justified. Consequently, it is not self-defeating for emotivists to appeal to evolutionary theory, which is a product of the capacity to form descriptive beliefs, to support emotivism.

Prescriptivists' evolutionary explanation holds that we have the capacity to issue moral commands because it increased the probability that our distant ancestors survived and reproduced. This explanation concerns the capacity to issue moral commands, but not the capacity to form descriptive beliefs. It does not at all indicate that no descriptive belief is justified. Accordingly, it is not self-defeating for prescriptivists to appeal to evolutionary theory to support prescriptivism.

Error theorists' evolutionary explanation holds that we have the capacity to form false beliefs because it was advantageous for our distant ancestors' survival and reproduction. This explanation concerns the capacity to form descriptive beliefs. In consequence, it is vulnerable to the charge that it is self-defeating for error theorists to appeal to evolutionary theory to support error theory. Avoiding this charge requires that error theorists should establish that the capacity to form moral beliefs is separate from the capacity to form descriptive beliefs, including evolutionary theory. Ideally, they also show that the former is completely unreliable while the latter is reliable.

Quasi-realists' evolutionary explanation holds that we have the capacity to have moral emotions because it increased our distant ancestors' viability and fertility. This explanation concerns the capacity to have moral emotions, but not the capacity to form descriptive beliefs. It does not at all indicate that no descriptive belief is justified. As a consequence, it is not self-defeating for quasi-realists to appeal to evolutionary theory to support quasi-realism.

Cultural emotivists' evolutionary explanation holds that we have the capacity to describe cultural emotions because it was an adaptive trait. It does not at all indicate that no descriptive belief is justified; it rather indicates that since some descriptive beliefs (beliefs about cultural emotions) are justified, other descriptive beliefs, including evolutionary theory, are also justified. In consequence, it is not self-defeating for cultural emotivists to appeal to evolutionary theory to support cultural emotivism.

Moral functionalists' evolutionary explanation holds that we have the capacity to encourage and discourage moral actions because it was an adaptive trait. This explanation concerns the capacity to regulate moral actions, but not the capacity to form descriptive beliefs. It does not at all indicate that no descriptive belief is justified. As a consequence, it is not self-defeating for moral functionalists to appeal to evolutionary theory to support cultural emotivism.

In sum, it is potentially self-defeating for moral skeptics and error theorists to appeal to evolutionary theory to justify moral skepticism and error theory, respectively. By contrast, it is not even potentially self-defeating for other metaethicists to appeal to evolutionary theory to justify their metaethical theories.

15.4.2 Maladaptive Traits

In this subsection, I argue that the explananda of moral skeptics' evolutionary explanation and error theorists' evolutionary explanation are maladaptive traits while those of other metaethicists' evolutionary explanations are adaptive traits.

In general, capacities to form justified beliefs are adaptive traits while capacities to form unjustified beliefs are maladaptive traits. This general claim about the relationship between the capacity to form beliefs and adaptiveness can be illustrated by the following imaginary example. Imagine that some Paleolithic individuals had the capacity, while other Paleolithic individuals did not, to form justified beliefs about who their siblings were and who their prospective partners were. It is obvious that the former would have been more likely to have healthy children than the latter.

Moral skepticism asserts that no moral judgment is justified. It implies that the capacity to make moral judgments is the capacity to form unjustified beliefs. However, the capacity to form unjustified beliefs is a maladaptive trait, given that in general, capacities to form unjustified beliefs are generally maladaptive traits. Therefore, moral skeptics cannot say that we have the capacity to make moral judgments because it was an adaptive trait. Furthermore, they have the burden of explaining why we have the maladaptive trait.

Moral skeptics might suggest that the capacity to make moral judgments differs from the capacity to form descriptive beliefs, such as the belief that a particular individual is not one's prospective partner but one's sibling. The difference is that the capacity to make moral judgments is an adaptive trait *even if* it produces unjustified beliefs, while the capacity to form descriptive beliefs is an adaptive trait *because* it produces justified beliefs. Therefore, they are right to say that we have the capacity to make moral judgments because it was an adaptive trait, and it is wrong for me to object that the capacity to make moral judgments is a maladaptive trait, and that moral skeptics owe us an explanation of why we have the maladaptive trait.

However, this move is ad hoc and calls for justification. Why apply a double standard to the capacity to make moral judgments and the capacity to form descriptive beliefs? Why is it that the capacity make moral judgments is an adaptive trait even if it produces unjustified beliefs, while the capacity to form descriptive beliefs is an adaptive trait because it produces justified beliefs? Why apply a low standard to the capacity to make moral judgments and a high standard to the capacity to form descriptive beliefs?

In the absence of an answer to this question, moral realists would apply the equal standard to the capacity to make moral judgments and to the capacity to form descriptive beliefs, saying that just as the capacity to form descriptive beliefs is an adaptive trait because it produces justified beliefs, so the capacity to make moral judgments is an adaptive trait because it produces justified beliefs. Moral skeptics would have the burden of showing that their ad hoc assumption is better than moral realists' ad hoc assumption.

Moral skeptics might now reject my previous contention that capacities to form justified beliefs are generally adaptive traits while capacities to form unjustified beliefs are generally maladaptive traits. They might suggest that we do not have any capacity to form justified beliefs, that all of our cognitive capacities produce unjustified beliefs but that all of them are adaptive traits. This suggestion involves global skepticism, according to which all of our beliefs in all domains are unjustified, i.e., none of them is justified.

An advantage of global skepticism is that it avoids my objection that it is ad hoc to say that the capacity to make moral judgments is an adaptive trait even if it produces unjustified beliefs, while the capacity to form descriptive beliefs is an adaptive trait because it produces justified beliefs. However, this advantage comes with a disadvantage: it is self-defeating to rely on evolutionary theory to establish global skepticism. After all, if none of our beliefs were justified, evolutionary theory would not be justified. If evolutionary theory were not justified, the debunking argument for global skepticism, which relies on the unjustified scientific theory, would not be cogent. As a result, global skepticism would collapse.

To summarize, moral skepticism holds that no moral belief is justified. It presupposes that the capacity to make moral judgments is the capacity to form unjustified beliefs. However, the capacity to form unjustified beliefs is a maladaptive trait, and hence moral skeptics cannot say that we have the capacity to make moral judgments because it was an adaptive trait, and they have the burden of explaining why we have the maladaptive trait.

The same objection applies mutatis mutandis to error theory. Error theory holds that no moral belief is true. It presupposes that the capacity to make moral judgments is the capacity to form false beliefs. However, the capacity to form false beliefs is a maladaptive trait. Just imagine what would have happened to the Paleolithic individuals who had false beliefs about who their siblings were and who their prospective partners were. Therefore, it is problematic for error theorists to suggest that we have the capacity to form false beliefs because it was an adaptive trait, and they owe us an explanation of why we have the maladaptive trait.

Similar objections do not apply to other metaethical explanations because none of them claims that the capacity to make moral judgments is the capacity to form moral beliefs, which are unjustified or false. For moral realists, the capacity to make moral judgments is the capacity to form beliefs, some of which are justified. For emotivists, the capacity to make moral judgments is the capacity to have moral emotions. For prescriptivists, the capacity to make moral judgments is the capacity to issue moral commands. For quasi-realists, the capacity to make moral judgments is the capacity to have moral emotions. For cultural emotivists, the capacity to make moral judgments is the capacity to form beliefs about cultural emotions. For moral functionalists, the capacity to make moral judgments is the capacity to influence moral actions. None of these capacities is a maladaptive trait.

15.4.3 A Moral Judgment and Its Opposite Moral Judgment

In this subsection, I argue that moral skepticism and error theory lack theoretical resources to distinguish between a moral judgment and its opposite moral judgment, while other metaethical theories do not.

As noted in Sect. 12.3 of Chap. 12, mathematical fictionalism asserts that no mathematical belief is true, and thus that it lacks the theoretical resource to distinguish between what laypeople take to be true mathematical statements and what they take to be false mathematical statements. For example, what is the difference between 3 + 3 = 6 and 3 + 3 = 9? Mathematical fictionalism does not have an adequate answer to this question.

Similarly, error theory asserts that no moral belief is true, and thus it lacks the theoretical resource to distinguish between a moral judgment and its opposite moral judgment. For example, what is the difference between the moral judgment that donating blood is right and its opposite moral judgment that donating blood is wrong? Error theory does not have an answer to this question (Sect. 12.3 of Chap. 12).

A similar problem plagues moral skepticism. Moral skepticism asserts that no moral judgment is justified, which implies that the moral judgment that donating blood is right is not justified, and that its opposite moral judgment that donating blood is wrong is not justified either. However, it sounds counterintuitive to say that there is no difference between the two moral judgments.

Moral skeptics might reply that there is a difference between the two moral judgments. The difference between them is that the moral judgment that donating blood is right increases our viability and fertility while its opposite moral judgment that donating blood is wrong decreases our viability and fertility. To put it in general terms, the difference between a moral judgment and its opposite moral judgment from the moral skeptical point of view is that one increases fitness while the other decreases it.

This possible suggestion has the following problem. The notion of fitness is external to moral skepticism, i.e., the notion of fitness is not part of moral skepticism, according to which no moral judgment is justified. It follows that moral realists as well as moral skeptics can avail themselves of the possible suggestion that a difference between a moral judgment and its opposite moral judgment is that one increases fitness while the other decreases it.

Unlike moral skepticism, moral realism has the theoretical resource to distinguish between a moral judgment and its opposite judgment. On the moral realist account, one is true while the other is false. For example, it is true that donating blood is right while it is false that donating blood is wrong. The notion of moral truth is internal to moral realism, and it is not available to moral skeptics who claim that no moral judgment is justified. Therefore, unlike moral skepticism, moral realism has the sui generis theoretical resource to distinguish between a moral judgment and its opposite moral judgment.

Like moral realism, emotivism has the sui generis theoretical resource to distinguish between a moral judgment and its opposite moral judgment. On the emotivist account, one is a positive emotion toward an action while the other is a negative emotion toward it. For example, the moral judgment that donating blood is right is a positive emotion toward donating blood, while its opposite moral judgment that donating blood is wrong is a negative emotion toward donating blood.

Like emotivism, other metaethical theories have the sui generis theoretical resources to distinguish between a moral judgment and its opposite moral judgment. On the prescriptivist account, one is a positive command about an action while the other is a negative command about it. On the quasi-realist account, one is a positive emotion toward an action while the other is a negative emotion toward it. On the cultural emotivist account, one is true of a cultural emotion while the other is false of it. On the moral functionalist account, one encourages some actions while the other discourages them.

15.4.4 The Principle of Charity

In this subsection, I argue that moral skepticism and error theory violate the principle of charity, just as mathematical fictionalism (Balaguer, 2018) does, but other metaethical theories do not.

According to mathematical fictionalism, no mathematical fact exists, and thus no mathematical belief is true. Since mathematical fictionalists believe that no mathematical belief is true, they cannot but attribute false beliefs to those hold mathematical beliefs. However, the principle of charity requires that we should assume ceteris paribus that others are rational agents, and thus that other things being equal, we should not attribute false beliefs to them. Mathematical fictionalists violate the principle of charity (Park, 2017, p. 79).

Just as mathematical fictionalists violate the principle of charity, so error theorists violate the principle of charity, as noted in Sect. 12.3 of Chap. 12. Error theorists attribute only false beliefs to moral evaluators, whether moral evaluators claim that an action is right or wrong. On this count, error theorists are no better off than mathematical fictionalists.

A similar problem arises for moral skeptics. The principle of charity also requires that other things being equal, we should not attribute unjustified beliefs to others. However, moral skeptics attribute only unjustified beliefs to moral evaluators, no matter how strong arguments they may construct for their moral views. As a result, they violate the principle of charity.

In contrast to moral skeptics and error theorists, other metaethical theorists do not violate the principle of charity. None of moral realists, emotivists, prescriptivists, quasi-realists, cultural emotivists, and moral functionalists claims that no moral belief is true, or that no moral belief is justified. It follows that none of them attributes false beliefs or unjustified beliefs to moral evaluators.

15.4.5 The Skeptic's Paradox

In this subsection, I argue that a paradox might arise when moral skeptics and error theorists express moral judgments, but not when other metaethicists express moral judgments.

Let me first clarify a paradox called "Moore's paradox." It arises when we assert a sentence of the form, "P, but I believe not p" or "P, but I don't believe p" (Moore, 1993, pp. 207–212). Imagine that a witness testifies in court, "The defendant killed the victim, but I believe it is false the defendant killed the victim." This speech act would puzzle listeners, including judges and prosecutors, and it would not lead to the conviction of the defendant (Park, 2020, p. 42). Listeners might even take it to be a symptom of irrationality.

What if the witness says, "The defendant killed the victim, but my belief that the defendant killed the victim is not justified"? This new speech act involves a similar paradox that I call "the skeptic's paradox." To put it in general terms, the skeptic's paradox arises when skeptics of p assert a sentence of the form, "P, but my belief that p is not justified," where p might be a moral proposition or a descriptive proposition This new paradox is so named because skeptics, those who believe that p is not justified, would be caught in it.

Moral skeptics claim that no moral judgment is justified. It follows that they can be caught in the skeptic's paradox. For example, they would be caught in it when they say, "Torture is wrong, but my belief that torture is wrong is not justified." Their speech act would puzzle their audience, and their audience might even take it to be a symptom of irrationality.

Moral skeptics can avoid the skeptic's paradox, provided that they do not add, "But my belief that torture is wrong is not justified," after saying, "Torture is wrong." In such a case, however, a new problem might arise, i.e., their audience might ask them the following disconcerting questions. "Don't you think your belief torture is wrong is not justified?" "How can you say something to me you take to be unjustified?" "Do you want me to believe something you take to be unjustified?" "How would you feel if I say something to you I take to be unjustified?" These questions constitute a philosophical problem called "the problem of disconcerting questions."

The problem of disconcerting questions also arises when you similarly attempt to avoid Moore's paradox. Suppose that you say, "The defendant killed the victim," but do not add, "But I believe it is false that the defendant killed the victim." Your audience might ask you the following disconcerting questions. "Don't you think it is false the defendant killed the victim?" "How can you say something to me that you take to be false?" "Do you want me to believe something you take to be false?" "How would you feel if I say something to you I take to be false?" Your audience might even think that your speech act is (morally) wrong (Park, 2019, p. 155).

Moral skeptics can get around the problem of disconcerting questions, provided that they do not express moral judgments. If they do not say, "Torture is wrong," no one will ask them the disconcerting questions. However, this way of overcoming the problem of disconcerting questions is costly. If none of us expresses moral judgments, our society will fall apart.

Error theorists might run into both Moore's paradox and the skeptic's paradox. After error theorists say, "Torture is wrong," they have to add, "But I believe it is false that torture is wrong" in line with error theory, thereby being caught in Moore's paradox. They also have to add, "But my belief that torture is wrong is not justified" in conformity with error theory, thereby being caught in the skeptic's paradox.

In contrast to moral skepticism and error theory, other metaethical theories do not claim that no belief is justified, and hence the skeptic's paradox does not arise for them. None of moral realism, emotivism, prescriptivism, quasi-realism, cultural emotivism, and moral functionalism claims that no moral belief is justified. The proponents of these metathetical theories would not add, "But my belief that donating blood is right is not justified," after saying, "Donating blood is right." It follows that their metaethical theories are pragmatically better than are moral skepticism and error theory.

15.4.6 Why Construct Arguments?

In this subsection, I argue that moral skepticism and error theory cannot explain, while other metaethical theories can, why applied and normative ethicists construct arguments.

Applied ethics is an attempt to classify actions as right and wrong. For example, omnivorians and vegetarians classify eating meat as right and wrong, respectively. Omnivorians argue that just as nonhuman omnivores have the right to eat meat, so humans have the right to eat meat. By contrast, vegetarians argue that just as cannibalism is wrong, so eating meat is wrong (Sect. 10.5 of Chap. 10). In short, applied ethicists construct arguments to justify their classifications of actions as right or wrong.

Normative ethics is an attempt to explain why some actions are right or wrong, and a related attempt to provide criteria for distinguishing between right and wrong actions. For example, utilitarianism asserts that charitable giving is right because it maximizes happiness, and that the maximization of happiness is the criterion for distinguishing between right and wrong actions. Other normative theories, such as Kantianism, cultural relativism, and divine command theory, provide different explanations of why some actions are right or wrong, and different criteria for distinguishing between right and wrong actions.

Normative ethicists construct arguments for and against rival explanations of why some actions are right or wrong, and for and against rival criteria for distinguishing between right and wrong actions. For example, critics of utilitarianism argue that it maximizes happiness to save six patients by pulling out organs from a healthy person against his will, but it is wrong to do so. As a consequence, utilitarianism cannot explain why it is wrong to do so, and the maximization of happiness is not a tenable criterion for distinguishing between right and wrong actions.

Utilitarians reply that you should explore whether there is an alternative way to save six patients before you pull out organs from a healthy person against his will.

In my view, moral skepticism cannot explain why applied ethicists construct arguments for and against moral beliefs. On the moral skeptical account, no moral belief is justified, and thus it is futile for omnivorians and vegetarians to construct arguments for and against omnivorism and vegetarianism, respectively. Omnivorians and vegetarians have the capacity to form moral beliefs, including the ones about eating meat, not because the capacity to form moral beliefs tracked moral facts but because it was fitness-conducive. Accordingly, neither omnivorism nor vegetarianism is justified, no matter how strong arguments omnivorians and vegetarians might come up with. The evolutionary argument trumps any argument in favor of omnivorism or vegetarianism. Note that moral skepticism cannot explain why applied ethicists construct arguments.

Neither can moral skepticism explain why normative ethicists construct arguments for and against rival explanations of, say, why charitable giving is right and why torture is wrong. Rival explanations of the moral explananda presuppose that the moral explananda are real, i.e., that charitable giving has the property of rightness, and that torture has the property of wrongness. On the moral skeptical account, however, we are justified in believing neither that charitable giving has the property of rightness nor that torture has the property of wrongness, although we might be justified in believing that charitable giving and torture exist. It is a waste of time for normative ethicists to explain why charitable giving is right and why torture is wrong. In general, if we are not justified in believing that something exists, it is pointless to explain why it exists. So moral skeptics would argue. Note that moral skepticism cannot explain why normative ethicists construct arguments for and against rival explanations of why some actions are right or wrong.

Neither can moral skepticism explain why normative ethicists construct arguments for and against rival criteria for distinguishing between right and wrong actions. Normative ethicists presuppose that some actions have the property of rightness while other actions have the property of wrongness. On the moral skeptical account, however, we are justified in believing neither that some actions have the property of rightness nor that other actions have the property of wrongness, although we might be justified in believing that actions exist. It is in vain for normative ethicists to provide rival criteria for distinguishing between some actions that allegedly have the property of rightness and other actions that allegedly have the property of wrongness. In general, if we are not justified in believing that X and Y exist, it is pointless to provide criteria for distinguishing between X and Y. Note that moral skepticism cannot explain why normative ethicists construct arguments for and against a criterion for distinguishing between right and wrong actions.

Like moral skepticism, error theory cannot explain why applied and normative ethicists construct arguments. On error theory, it is false that eating meat is right, and it is also false that eating meat is wrong. It is mysterious why omnivorians and vegetarians construct arguments for and against false beliefs. It is also mysterious why normative ethicists construct arguments for and against normative ethical theories. Normative ethicists think that their theories explain why some actions are right or wrong, and provide rival criteria for distinguishing between right and wrong actions. However, no action has a moral property of rightness or wrongness. Therefore, it is futile to explain why some actions are right or wrong, and to provide a criterion for distinguishing between right and wrong actions. Note that error theory cannot explain why applied and normative ethicists construct arguments.

In contrast to moral skepticism and error theory, other metaethical theories can explain why applied ethicists construct arguments for and against moral beliefs. For example, why do omnivorians and vegetarians construct arguments for and against omnivorism and vegetarianism? On moral realism, the debate between them helps us to choose between the belief that eating meat has the property of rightness and the belief that eating meat has the property of wrongness[3] (Park, 2012, p. 184). On emotivism and quasi-realism, the debate helps us to choose between a positive emotion and a negative emotion toward eating meat (Park, 2012, p. 183). On prescriptivism, the debate helps us to choose between a positive command and a negative command about eating meat. On cultural emotivism, the debate helps our culture to choose between a positive emotion and a negative emotion toward eating meat, and also helps us to choose between the belief that our culture has a positive emotion and the belief that our culture has a negative emotion toward eating meat. On moral functionalism, the debate helps us to choose between encouraging and discouraging eating meat.

Other metaethical theories can also explain why normative ethicists construct arguments for and against rival explanations of why some actions are right or wrong. For example, rival normative ethicists debate about rival explanations of why torture is wrong. On moral realism, the debate between them helps us to understand why torture has the property of wrongness. On emotivism and quasi-realism, the debate helps us to understand why we have a negative emotion toward torture. On prescriptivism, the debate helps us to understand why we have a negative command about torture. On cultural emotivism, the debate helps us to understand why our culture has a negative emotion toward torture, and also why we believe that our culture has a negative emotion toward torture. On moral functionalism, the debate helps us to understand why we discourage torture.

Other metaethical theories can also explain why normative ethicists construct arguments for and against rival criteria for distinguishing between right and wrong actions. On moral realism, a criterion helps us to distinguish between actions that have the property of rightness and actions that have the property of wrongness. On emotivism and quasi-realism, a criterion helps us to choose between a positive emotion and a negative emotion toward certain actions. On prescriptivism, a criterion helps us to choose between a positive command and a negative command about certain actions. On cultural emotivism, a criterion helps us to form the belief about whether our culture has a positive emotion or a negative emotion toward certain actions. On moral functionalism, a criterion helps us to choose between encouraging and discouraging certain actions.

[3] See Sect. 10.6 of Chap. 10 for my objections to this suggestion.

So far, I have argued that moral realism, emotivism, prescriptivism, quasi-realism, cultural emotivism, and moral functionalism can explain, while moral skepticism and error theory cannot, why applied ethicists construct arguments for and against moral beliefs, why normative ethicists construct arguments for and against rival explanations of why some actions are right or wrong, and why normative ethicists construct arguments for and against rival criteria for distinguishing between right and wrong actions.

So what? Moral skeptics are in the following quagmire. At the end of Sect. 15.3 above, I argued that they have the burden of demonstrating that their evolutionary explanation is better than other metaethicists' evolutionary explanations. They would have to construct an argument to meet that burden. An interesting issue is whether it is justifiable that the act of constructing an argument is explainable in metaethics, but not in applied and normative ethics. It sounds ad hoc for moral skeptics to say that metaethics is different from applied and normative ethics such that metaethicists' constructing an argument is explainable, but applied and normative ethics' constructing arguments is not.

How would moral skeptics and error theorists respond to my objection that they cannot explain, while other metaethicists can, why applied and normative ethicists construct arguments? They might criticize other metaethicists' explanations of why applied and normative ethicists construct arguments, and then claim that none of them is tenable. Hence, moral skepticism and error theory are no worse off than other metaethical theories.

My response to this possible move is to say that moral skeptics and error theorists' criticisms against the metaethical explanations can serve as the starting point of an interesting debate among rival metaethicists. Suffice it to say here that providing the metaethical explanations is better than providing no metaethical explanation at all. As James Rachels and Stuart Rachels put it, "A moral judgment—or for that matter, any kind of value judgment—must be supported by good reasons. Any adequate theory of ethics should be able to explain how reason can support moral judgments" (Rachels & Rachels, 2010, pp. 39–40). By Rachels and Rachels's standard, moral skepticism and error theory are worse than other metaethical theories.

15.5 Conclusion

Moral skeptics' evolutionary explanation of why we have the capacity to make moral judgments is no less parsimonious than other metaethical evolutionary explanations provided by emotivists, prescriptivists, quasi-realists, and moral functionalists, given that none of the other metaethical evolutionary explanations invokes moral facts. Moral realism and cultural emotivism have wider explanatory scopes than moral skepticism, given that moral realism and cultural emotivism explain, while moral skepticism does not, why the capacity to make moral judgments was an adaptive trait. None of the six objections to moral skepticism and error theory applies to moral realism, emotivism, prescriptivism, quasi-realism, cultural emotivism, or moral functionalism. This chapter has the following two messages for moral skeptics and error theorists: "Moral skeptics concede too much to cognitivists" and "Moral skepticism and error theory are less promising than other metaethical theories."

References

Balaguer, M. (2018). Fictionalism in the philosophy of mathematics. In E. N. Zalta (Ed.), *The Stanford encyclopedia of philosophy*. https://plato.stanford.edu/archives/fall2018/entries/fictionalism-mathematics/

Blackburn, S. (1993). *Essays in quasi-realism*. Oxford University Press.

De Cruz, H., Boudry, M., De Smedt, J., & Blancke, S. (2011). Evolutionary approaches to epistemic justification. *Dialectica, 65*(4), 517–535.

Goldman, A., & Beddor, B. (2021). Reliabilist epistemology. In E. N. Zalta (Ed.), *The Stanford encyclopedia of philosophy*. https://plato.stanford.edu/archives/sum2021/entries/reliabilism/

Hare, R. (1952). *Language of morals*. Clarendon Press.

Mackie, J. (1977). *Ethics: Inventing right and wrong*. Penguin Books.

Moore, G. (1993). Moore's paradox. In T. Baldwin (Ed.), *G.E. Moore: Selected writings* (pp. 207–212). Routledge.

Park, S. (2012). Against moral truths. *Cultura: International Journal of Philosophy of Culture and Axiology, 9*(1), 179–194.

Park, S. (2017). In defense of mathematical inferentialism. *Analysis and Metaphysics, 16*, 70–83.

Park, S. (2019). Should scientists embrace scientific realism or antirealism? *Philosophical Forum, 50*(1), 147–158.

Park, S. (2020). Replies to Healey's comments regarding van Fraassen's positions. *Social Epistemology Review and Reply Collective, 9*(1), 38–47.

Rachels, J., & Rachels, S. (2010). *The elements of moral philosophy*. McGraw-Hill Humanities.

Stevenson, C. (1944). *Ethics and language*. Yale University Press.

Street, S. (2006). A Darwinian dilemma for realist theories of value. *Philosophical Studies, 127*(1), 109–166.

Chapter 16
The Problem of Moral Induction

Abstract Just as Hume constructs an argument against factual inductions, so I construct a Humean argument against moral inductions. Just as Hume's argument against factual inductions creates the problem of factual induction, so the Humean argument against moral inductions creates the problem of moral induction. The problem of factual induction spells trouble for the evolutionary debunking argument for moral skepticism. The problem of moral induction spells trouble for moral realism and cultural emotivism, but not for emotivism, prescriptivism, error theory, quasi-realism, and moral functionalism. The problem of moral induction can trigger new meta-normative debates.

Keywords Moral uniformity principle · Problem of induction · Problem of moral induction · Uniformity principle

16.1 Introduction

It is a factual inductive inference that since observed pine trees are green, unobserved pine trees are also green. It is also a factual inductive inference that since overeating and obesity were conjoined, they will also be conjoined. David Hume (1993, pp. 15–25) constructs an argument against factual inductions, which results in the famous philosophical problem in the history of philosophy called "the problem of induction" according to which factual inductions, inferences from factual statements about observed objects to factual statements about unobserved objects, are not justifiable.

It is a moral inductive inference, for example, that since observed theft is (morally) wrong, unobserved theft is also wrong. Can we construct an argument against moral inductions similar to Hume's argument against factual inductions? My answer is "Yes." Just as Hume's argument against factual inductions results in the problem of factual induction, so the argument against moral inductions results in the problem of moral induction, according to which moral inductions, inferences from moral

S. Park, *Induction, Science, and Morality*, Synthese Library 520,
https://doi.org/10.1007/978-3-032-00032-3_16

statements about observed actions to moral statements about unobserved actions, are not justifiable.

This chapter is organized as follows. In Sect. 16.2, I briefly sketch Hume's argument against factual inductions and its implication for the evolutionary debunking argument for moral skepticism (Street, 2006). In Sect. 16.3, I spell out the Humean argument against moral inductions and its implication for moral realism. In Sect. 16.4, I argue that moral inductive skepticism has several disadvantages. In Sect. 16.5, I criticize the Humean argument against moral inductions in the way I criticized Hume's argument against factual inductions in Sect. 1.3 of Chap. 1 and in Sect. 5.3 of Chap. 5. In Sect. 16.6, I argue that the problem of moral induction undercuts cultural emotivism, but not emotivism, prescriptivism, error theory, quasi-realism, and moral functionalism. In Sect. 16.7, I sketch new debates that meta-normative theorists can be engaged in once they take Hume's problem of factual induction seriously.

This chapter is intended to be useful to metaethicists who wonder how Hume's problem of factual induction relates to metaethical theories, and to meta-normative theorists who aim to develop new debates over new meta-normative topics.

16.2 Hume's Argument Against Factual Inductions

Why is Hume skeptical about factual inductions? Why does he think that factual inductions are not justifiable? In this section, I briefly sketch his argument against factual inductions and then its disastrous implication for the debunking argument for moral skepticism (Street, 2006).

Hume's argument for inductive skepticism can be illustrated with the use of the following example of an inductive argument:

(P)
Observed pine trees are green.
∴ Unobserved pine trees are green.

On Hume's account, the premise of (P) is justified because we have empirical evidence for it. By contrast, the conclusion of (P) is neither justified a priori nor a posteriori. It is not justified a priori because negating it does not give rise to a contradiction. It is not justified a posteriori either because there is no necessary connection between the property of being a pine tree and the property of being green (Sect. 1.2 of Chap. 1).

Hume claims that the uniformity principle, which holds that nature is uniform, is required for the inductive inference of (P) to be justified. If the uniformity principle were justified, it could be used, along with the premise, to justify the conclusion of (P). To state it otherwise, the premise alone does not justify the conclusion, but it does in cooperation with the uniformity principle. However, the uniformity principle is neither justified a priori nor a posteriori. It is not justified a priori because a contradiction does not arise from negating it. It is not justified a posteriori either because it is circular to appeal to inductive arguments, such as (P), to justify it. Therefore, (P) is not justifiable (Sect. 1.2 of Chap. 1).

How does Hume's argument against factual inductions relate to the debunking argument for moral skepticism (Street, 2006)? The debunking argument is basically composed of the premise that we have the capacity to form moral beliefs because it was an adaptive trait, and the conclusion that no moral belief is justified (Sect. 15.2 of Chap. 15).

Hume's argument against factual inductions has a disastrous consequence on the premise of the debunking argument. The premise involves the inference from the explanandum that we have the capacity to form moral beliefs to the explanans that it was an adaptive trait. The inference is inductive, given that even if the explanandum is true, the explanans might be false. Since the inference is inductive, it falls victim to Hume's argument against inductions, i.e., it is not justifiable.

Hume's argument against factual inductions also has a disastrous consequence on the inference from the premise to the conclusion of the debunking argument. The inference is inductive, given that no contradiction arises from affirming the premise that we have the capacity to form them because it was an adaptive trait, but denying the conclusion that no moral belief is justified. Since the inference is inductive, it falls prey to Hume's argument against inductions. As a result, it is questionable whether no moral belief is justified.

So what? Debunkers have to take Hume's argument against factual inductions seriously and attempt to refute it. Debunkers are alone on this count, as will become clear in the next section.

16.3 The Humean Argument Against Moral Inductions

In this section, I construct a Humean argument against moral inductions similar to Hume's argument against factual inductions in Sect. 16.2 above, and elucidate its disastrous implication for moral realism.

The Humean argument against moral inductions can be illustrated with the use of the following example of a moral induction:

(T)
Observed theft is wrong.
… Unobserved theft is wrong.

Hume would say that the premise of (T) is neither justified a priori nor a posteriori. It is not justified a priori because no contradiction is involved in its negation. It is not justified a posteriori either because moral properties are imperceptible (Hume, 1978, pp. 468–469).

However, we might grant for the sake of argument that the premise of (T) is justified a posteriori,[1] and then explore whether the conclusion of (T) is justified. Moral inductive skeptics, those who are skeptical about the conclusions of moral inductions, would say that the conclusion of (T) is neither justified a priori nor a posteriori. It is not justified a priori because its negation does not commit a contradiction. It is not justified a posteriori either because there is no necessary connection between theft and wrongness.

Moral inductive skeptics would claim that a general principle is required for the conclusion of (T) to be justified. I call it "the moral uniformity principle" according to which morality is uniform. If it were justified, it could be used, along with the premise of (T), to justify the conclusion of (T). To state it otherwise, the premise of (T) alone does not justify the conclusion of (T), but it does in cooperation with the moral uniformity principle.

Moral inductive skeptics would argue that the moral uniformity principle is not justified a priori because a contradiction does not arise from negating it. It is not justified a posteriori either because it is circular to appeal to moral inductions, such as (T), to justify the moral uniformity principle. Since the moral uniformity principle is neither justified a priori nor a posteriori, the conclusion of (T) is not justifiable.

The Humean argument against moral inductions recounted so far can be summed up in the following standard form:

Moral inductions are neither justified a priori nor a posteriori.
The moral uniformity principle is required for moral inductions to be justified.
The moral uniformity principle is neither justified a priori nor a posteriori.
∴ Moral inductions are not justifiable.

This argument has a disastrous implication for moral realism. Moral realism is the view that we know that some moral statements are true (Sect. 10.2 of Chap. 10). Moral realists believe that there is no fundamental difference between descriptive and moral statements. Just as descriptive statements are true or false, depending on whether they match up with descriptive facts or not, so moral statements are true or false, depending on whether they match up with moral facts or not.

Moral realists make the inductive inference from the premise that observed theft is wrong to the conclusion that unobserved theft is wrong. Moral inductive skeptics would object that the conclusion is not justifiable. Neither the faculty of reason nor experience justifies the conclusion. The moral uniformity principle is required for the conclusion to be justified, but neither the faculty of reason nor experience justifies the moral uniformity principle. Accordingly, the conclusion is not justifiable.

[1] Andrew Cullison (2010) claims that moral properties are perceptible.

16.4 The Disadvantages of Moral Inductive Skepticism

In this section, I argue that moral inductive skepticism has several disadvantages.

The Humean argument against moral inductions leads us to moral inductive skepticism, as noted in Sect. 16.3 above. In general, skepticism has several pragmatic and epistemic disadvantages (Park, 2019, pp. 96–97). This point applies to inductive moral skepticism. For example, moral inductive skepticism cannot explain any moral phenomenon. After all, to explain a moral explanandum in terms of a moral explanans involves the inductive inference from the moral explanandum to the moral explanans. Moral inductive skeptics are skeptical about such inferences. Therefore, they cannot explain anything. In short, "Believe little. Explain little" (Park, 2015, p. 28).

This disadvantage with moral skepticism can be illustrated by the following three examples.

(i) Some moral agents give to charity. Why do they give to charity? On the moral realist account, they give to charity because giving to charity has the property of rightness (Sect. 10.3 of Chap. 10). Such an explanation involves the inductive inference from the observation of some moral agents' behavior to the existence of the moral property or to the general belief that giving to charity is right. Such inferences are questionable according to moral inductive skepticism. In consequence, moral inductive skeptics cannot explain why some moral agents give to charity.

(ii) The capacity to make moral judgments was fitness-enhancing (Sect. 15.2 of Chap. 15). Why was it fitness-enhancing? On the moral realist account, it was fitness-enhancing because it tracked moral facts. This explanation involves the inductive inference from the explanandum that the capacity to make moral judgments was fitness-enhancing to the explanans that it tracked moral facts. Such inferences are dubious according to moral inductive skepticism. As a consequence, moral inductive skeptics cannot explain why the capacity to make moral judgments was fitness-enhancing.

(iii) Applied and normative ethicists construct arguments (Sect. 15.4.6 of Chap. 15). Why do they construct arguments? On the moral realist account, applied ethicists construct arguments because constructing arguments helps moral evaluators to see a moral property in an action that applied ethicists think that they see, and normative ethicists construct arguments because constructing arguments helps moral evaluators to understand why an action has the property of rightness or wrongness, and to acquire a criterion for distinguishing between right and wrong actions. These explanations involve the inductive inferences from the explananda to the explanantia. Such inferences are distrustful according to moral inductive skepticism. Accordingly, moral inductive skeptics cannot explain why applied and normative ethicists construct arguments.

Moral inductive skepticism has another disadvantage that is worth mentioning here. It runs into the skeptic's paradox. The skeptic's paradox arises when skeptics of p assert a sentence of the form "P, but my belief that p is not justified" (Sect. 15.4.5 of Chap. 15). For example, imagine that a witness testifies in court, "The accused robbed the bank, but my belief that the accused robbed the bank is not justified." Such a speech act would puzzle the witness's listeners, including judges and prosecutors. As a result, it would not lead to the conviction of the accused, and it can even be taken to be a symptom of irrationality. Similarly, the skeptic's paradox arises when a moral inductive skeptic says, "Theft is wrong, but my belief that theft is wrong is not justified." Such a speech act would puzzle her listeners, and it can even be taken to be a symptom of irrationality (Sect. 15.4.5 of Chap. 15).

In sum, it is costly to embrace moral inductive skepticism. Those who resist the cost of embracing moral inductive skepticism have the burden of refuting the Humean argument against moral inductions. I offer some objections to it in the next section.

16.5 Critiques of the Humean Argument Against Moral Inductions

In this section, I argue that my objections to Hume's argument against factual inductions apply to the Humean argument against moral inductions.

In Sect. 5.3 of Chap. 5, we have seen that the process of generating the problem of factual induction fits the template for generating philosophical problems, such as the problem of simplicity, the problem of generalizability, the argument from a bad lot (van Fraassen, 1989, p. 143), and the argument from a brute fact (de Ray, 2022). Does the process of generating the problem of moral induction also fit the template? In other words, can the template be also used to generate the problem of moral induction?

My answer is, "Yes." The use of the template consists of the following three steps. First, moral inductive skeptics point out that moral beliefs about unobserved actions might be false, arguing that the negations of moral beliefs about unobserved actions are not contradictions, and that actions and moral properties are not necessarily connected with each other. Second, moral inductive skeptics suggest that the moral uniformity principle is required for moral beliefs about unobserved actions to be justified. Third, moral inductive skeptics claim that no moral belief about unobserved actions is justified, and that those who hold moral beliefs about unobserved actions have the burden of justifying the moral uniformity principle without relying on a moral induction. Going through these three steps result in the problem of moral induction.

To say that the process of generating a philosophical problem fits the template means that it is a product of confounding a means and an end (Sect. 5.3 of Chap. 5). This point applies to the problem of moral induction. Moral inductions are the means to arrive at the end of the moral uniformity principle rather than the other way around. Therefore, it is accurate to say that moral inductions are required for the moral uniformity principle to be justified, but inaccurate to say that the moral uniformity principle is required for moral inductions to be justified.

My other objections to Hume's argument against inductions in Sect. 1.3 of Chap. 1 can be leveled at the Humean argument against moral inductions. For example, it is an inductive inference that since the conclusions of moral inductions are neither justified a priori nor a posteriori, they are not justified. It is also an inductive inference that since the moral uniformity principle is neither justified a priori nor a posteriori, it is not justified. It is self-defeating for moral inductive skeptics to rely on such inferences to show that moral inductions are not justifiable.

Moral realists who accept my objections to the Humean argument against moral inductions would also have to accept my objections to Hume's argument against factual inductions in Sect. 1.3 of Chap. 1. It would be arbitrary for moral realists to say that my objections to the Humean argument against moral inductions are acceptable while my objections to Hume's argument against factual inductions are not.

16.6 Other Metaethical Theories

In this section, I argue that the problem of moral induction spells trouble for cultural emotivism, but not for emotivism, prescriptivism, error theory, quasi-realism, and moral functionalism.

The Humean argument against moral inductions presupposes that the premises and the conclusions of moral inductions express beliefs, and thus that they are truth-evaluable and can be shown to be likely to be true by evidence. Thus, cognitivism, the view that a moral judgment is a belief, is built into the Humean argument against moral inductions. Moral realism is a form of cognitivism. It is natural that it is vulnerable to the problem of moral induction.

Cultural emotivism is another form of cognitivism. It holds that a moral judgment is a belief about a cultural emotion (Sect. 10.3 of Chap. 10). Under the framework of cultural emotivism, inductive skeptics would point out that the Humean argument against moral inductions applies to the conclusion of (T). The conclusion of (T) is neither justified a priori nor a posteriori. The moral uniformity principle is required for the premise of (T) to justify the conclusion of (T) to be justified. However, it is neither justified a priori nor a posteriori. Therefore, the conclusion of (T) is not justifiable.

Let me turn to emotivism (Stevenson, 1944). Emotivism is not a form of cognitivism but a form of noncognitivism. It asserts that a moral statement is an expression of emotion. Under the framework of emotivism, moral inductive skeptics would say nothing about the conclusion of (T). After all, moral inductive skepticism applies to moral beliefs, but not to moral emotions. If moral inductive skeptics say that moral evaluators are not justified in believing that unobserved theft is wrong, emotivists will retort that an unjustified belief about unobserved theft is not attributable to moral evaluators, and that a negative emotion about unobserved theft is rather attributable to them.

Let me turn to prescriptivism (Hare, 1952). Like emotivism, prescriptivism is a form of noncognitivism. It asserts that a moral statement is an expression of a moral command. Under the framework of prescriptivism, moral inductive skeptics would be reticent about the conclusion of (T). After all, moral inductive skepticism applies to moral beliefs, but not to moral commands. If moral inductive skeptics say that moral evaluators are not justified in believing that unobserved theft is wrong, prescriptivists will retort that an unjustified belief about unobserved theft is not attributable to moral evaluators, and that a negative command about unobserved theft is rather attributable to them.

Let me turn to error theory (Mackie, 1977). Like moral realism, error theory is a form of cognitivism, asserting that a moral statement expresses a belief. However, it asserts that no moral fact exists, and thus that no moral belief is true. Error theorists would agree with inductive moral skeptics that the conclusion of (T) is not justified. However, they would disagree with moral inductive skeptics on the issue of whether it is false. They would say that it is false, while moral inductive skeptics would leave the possibility open whether it is true. Thus, error theory is a more radical metaethical theory than is moral inductive skepticism.

Let me move on to quasi-realism (Blackburn, 1993). Like emotivism, it is a form of noncognitivism. However, it allows for the use of the predicate "true" in moral discourse, asserting that to say that p is true is the same as to say that p. Under the framework of quasi-realism, moral inductive skeptics would be silent about the conclusion of (T), given that moral inductive skepticism does not apply to noncognitive states. If moral inductive skeptics say that moral evaluators are not justified in believing that unobserved theft is wrong, quasi-realists will retort that an unjustified belief about unobserved theft is not attributable to moral evaluators, and that a negative emotion toward unobserved theft is rather attributable to them.

Let me turn to moral functionalism. Moral functionalism asserts that moral utterances perform the functions of encouraging and discouraging moral actions, and it is noncommittal about whether they express cognitive states (Sects. 11.2 and 11.5 of Chap. 11). Under the framework of moral functionalism, moral inductive skeptics might or might not say that the conclusion of (T) is not justified. Suppose that they say that it is not justified. Moral functionalists would be noncommittal about whether moral inductive skeptics are right that it is not justified. However, they would insist that moral inductive skeptics should never fail to attribute the action of discouraging unobserved theft to moral evaluators.

It appears that moral realists and cultural emotivists should take the problem of moral induction seriously, but other metaethicists do not need to. In the next section, however, I show that all metaethicists may have to take the problem of moral induction seriously.

16.7 New Meta-Normative Debates

In this section, I explore how rival metaethicists would respond to the opposite normative views that rival attitude-takers hold concerning the uniformity principle and other principles in Sect. 5.3 of Chap. 5. I also sketch how new debates might unfold between rival attitude-takers, and between meta-normative theorists and attitude-takers.

Hume and inductivists hold opposite views about the uniformity principle. Their opposite views can be illustrated with the use of the following inductive argument:

(P)
Observed pine trees are green.
… Unobserved pine trees are green.

On Hume's view, the uniformity principle is required for the premise to justify the conclusion. On the inductivist view, by contrast, the uniformity principle is not required for the premise to justify the conclusion, and the premise alone justifies the conclusion (Sect. 1.3.2 of Chap. 1). These opposite views about the uniformity principle are not descriptive ones but normative ones. Hume's view is that inductivists, who believe that the conclusion of (P) is true, should justify the uniformity principle, while inductivists' view is that they do not need to.

Rival metaethicists would respond differently to the opposite normative views. Moral realists would say that the opposite normative views are opposite beliefs about the uniformity principle. Emotivists would say that they are opposite emotions toward the uniformity principle. Prescriptivists would say that they are opposite commands about the uniformity principle. Error theorists would say that none of them is true. Quasi-realists would say that they are opposite emotions toward the uniformity principle, but we can say that they are true or false. Moral skeptics would say that none of them is justified. Cultural emotivists would say that they are opposite beliefs about the cultural emotion toward the uniformity principle. Moral functionalists would say that they perform opposite functions of encouraging and discouraging the justification of the uniformity principle.

The existence of the rival meta-normative responses to the opposite normative views means the existence of the rival meta-normative frameworks under which Hume and inductivists can debate about the uniformity principle. Under the cultural emotivist framework, for example, Hume might argue that he believes that his culture has a positive emotion toward the uniformity principle because inductions are neither justified a priori nor a posteriori and because they are not justified without the uniformity principle. By contrast, inductivists might argue that they believe that their culture has a negative emotion toward it because inductions are justified even without it and because inductions are required for it to be justified rather than the other way around.

Under the moral functionalist framework, Hume might argue that he encourages the justification of the uniformity principle that does not appeal to inductions because inductions are neither justified a priori nor a posteriori and because they are not justified without the uniformity principle. By contrast, inductivists might argue that they discourage the justification of the uniformity principle that does not appeal to inductions because inductions are justified even without it and because inductions are required for it to be justified rather than the other way around.

Just as Hume and inductivists hold opposite views about the uniformity principle, so Bas van Fraassen (1989, p. 143) and I hold opposite views about the privilege principle. Van Fraassen contends that the privilege principle is required, while I contend that it is not, for inference to the best explanation (IBE) to be justified (Sect. 4.4.3 of Chap. 4). Rival philosophers would similarly hold opposite views about the explainability principle, the memorial principle, the reliability principle, the nonexistence principle, the simplicity principle, the transparency principle, and the generalizability principle in Sect. 5.3 of Chap. 5. The proponents would claim that the principles are required, while the opponents would claim that they are not, for particular beliefs to be justified.

Just as rival metaethicists respond differently to Hume's normative view and inductivists' normative view about the uniformity principle, so they would respond differently to the proponents' normative views and the opponents' normative views about the other principles above. For example, moral realists would say that both the proponents and the opponents express opposite beliefs about the other principles. By contrast, emotivists would say that rival attitude-takers express opposite emotions toward the other privilege principles.

Just as Hume and inductivists can debate about the uniformity principle under rival meta-normative frameworks, so rival attitude-takers can debate about the other principles under rival meta-normative frameworks. Under the cultural emotivist framework, for example, van Fraassen might argue that he believes that his culture has a positive emotion toward the privilege principle because the conclusions of IBE might be from a bad lot and because they are not justified without the privilege principle. By contrast, scientific realists might argue that they believe that their culture has a negative emotion toward the privilege principle because IBE is justified even without the privilege principle and because IBE is required for the privilege principle to be justified rather than the other way around.

Under the moral functionalist framework, van Fraassen might argue that he encourages the justification of the privilege principle that does not appeal to IBE because the conclusions of IBE might be from a bad lot and because they are not justified without the privilege principle. By contrast, scientific realists might argue that they discourage the justification of the privilege principle that does not appeal to IBE because IBE is justified even without the privilege principle and because IBE is required for the privilege principle to be justified rather than the other way around.

There can also be interesting new debates between metaethicists and attitude-takers of the aforementioned principles. For example, moral skeptics (debunkers), who believe that no moral belief is justified, would contend that it is not a justified belief that the uniformity principle is required for inductions to be justified, and thus that Hume would have to justify the unjustified belief before he wields the problem of factual induction against the debunking argument for moral skepticism, i.e., before he says that the premise of the debunking argument exemplifies IBE, that the inference from the premise to the conclusion of the debunking argument is inductive, and thus that none of the premise and the conclusion is justified.

Hume, who believes that no conclusion of factual inductions is justified, would reply that moral skeptics (debunkers) would have to establish the debunking argument for moral skepticism, which requires inductions, before they wield moral skepticism against his argument for inductive skepticism, i.e., before they say that it is not a justified belief that the uniformity principle is required for inductions to be justified.

The foregoing confrontation between Hume and moral skeptics raises the fundamental issue of which takes precedence over which. Does Hume's normative view about the uniformity principle take precedence over moral skeptics' meta-normative view about his normative view? Or does moral skeptics' meta-normative view about his normative view take precedence over Hume's normative view? Hume would say that his normative view is prior to moral skeptics' meta-normative view about his normative view, and thus that his argument against factual inductions has a disastrous consequence on the debunking argument. By contrast, moral skeptics would say that their meta-normative view about his normative view is prior to his normative view, and thus that the debunking argument has a disastrous consequence on his argument against factual inductions.

The foregoing debate between Hume and moral skeptics might be joined by emotivists. Emotivists, who have embraced inductivism, would say to Hume that their meta-normative view about his normative view is prior to his normative view, and hence that they do not need to accept his normative view, i.e., they do not need to have a positive emotion toward the uniformity principle. They would even express a negative emotion toward the uniformity principle, i.e., they would even say that the uniformity principle is not required for inductions to be justified.

Similarly, moral functionalists, who have embraced inductivism, would say to Hume that their meta-normative view about his normative view is prior to his normative view, and hence that they do not need to accept his normative view, i.e., they do not need to encourage the justification of the uniformity principle that does not appeal to inductions. They would even discourage the justification of the uniformity principle that does not appeal to inductions, i.e., they would even say that the uniformity principle is not required for inductions to be justified.

This possible debate between Hume and meta-normative theorists is intended to give the readers of this book an idea of how new meta-normative debates might unfold between meta-normative theorists and attitude-takers of the principles. I leave the task of developing and enriching the new debates to future researchers. The new debates will shed new light on the issue of whether the principles are required for particular beliefs to be justified, and on the issue of whether the rival metaethical theories are tenable. As a result, the new debates will benefit both meta-normative theorists and attitude-takers of the principles.

16.8 Conclusion

Hume's argument against factual inductions poses a threat to the premise and to the inference from the premise to the conclusion of the debunking argument for moral skepticism. The Humean argument against moral inductions poses a threat to moral realism and to cultural emotivism, but not to other metaethical theories.

Rival attitude-takers of the principles hold rival normative views about whether the principles are required for particular beliefs to be justified. Rival metaethicists would hold rival meta-normative views about rival attitude-takers' normative views, providing rival meta-normative frameworks under which rival attitude-takers can engage in new debates over the principles.

There can also be new debates between meta-normative theorists and attitude-takers of the principles. Meta-normative theorists would say that their meta-normative view takes precedence over attitude-takers' normative view, while attitude-takers would say that their normative view takes precedence over meta-normative theorists' normative view.

This chapter can be summed up as follows. Taking Hume's problem of factual induction seriously can be a starting point of interesting new debates over meta-normative topics.

References

Blackburn, S. (1993). *Essays in quasi-realism*. Oxford University Press.
Cullison, A. (2010). Moral perception. *European Journal of Philosophy, 18*(2), 159–175.
de Ray, C. (2022). A new epistemological case for theism. *Religious Studies, 58*(2), 379–400.
Hare, R. (1952). *Language of morals*. Clarendon Press.
Hume, D. (1978). *A treatise of human nature*. In L. A. Selby-Bigge & P. H. Nidditch (Eds.). Oxford University Press.
Hume, D. (1993). *An enquiry concerning human understanding*. Hackett Publishing Company.
Mackie, J. (1977). *Ethics: Inventing right and wrong*. Penguin Books.
Park, S. (2015). Explanatory failures of relative realism. *Epistemologia, 38*(1), 16–28.
Park, S. (2019). The disastrous implications of the 'English' view of rationality in a social world. *Social Epistemology, 33*(1), 88–99.
Stevenson, C. (1944). *Ethics and language*. Yale University Press.
Street, S. (2006). A Darwinian dilemma for realist theories of value. *Philosophical Studies, 127*(1), 109–166.
van Fraassen, B. (1989). *Laws and symmetry*. Oxford University Press.

Index

S. Park, *Induction, Science, and Morality*, Synthese Library 520,
https://doi.org/10.1007/978-3-032-00032-3

The manufacturer's authorised representative in the EU is Springer Nature Customer Service Centre GmbH, Europaplatz 3, 69115 Heidelberg, Germany. If you have any concerns regarding our products, please contact ProductSafety@springernature.com

Printed and bound by CPI Group (UK) Ltd, Croydon, CR0 4YY
30/09/2026
02225589-0011